T0135659

Augsburger Schriften zur Mathematik, Physik und Informatik

Band 4

herausgegeben von:
Professor Dr. F. Pukelsheim
Professor Dr. B. Aulbach
Professor Dr. W. Reif
Professor Dr. D. Vollhardt

Bibliografische Information Der Deutschen Bibliothek

Die Deutsche Bibliothek verzeichnet diese Publikation in der Deutschen Nationalbibliografie; detaillierte bibliografische Daten sind im Internet über http://dnb.ddb.de abrufbar.

ISBN 3-8325-0530-X
ISSN 1611-4256
Logos Verlag Berlin
Comeniushof, Gubener Str. 47,
10243 Berlin
Tel.: +49 030 42 85 10 90

The Interplay of Structural and Electronic Properties in Transition Metal Oxides

Zur Erlangung des akademischen Grades eines
Doktors der Naturwissenschaften
der Mathematisch-Naturwissenschaftlichen Fakultät
der Universität Augsburg
vorgelegte

Dissertation

von

Dipl.-Phys. Udo Schwingenschlögl

aus Bobingen

Augsburg 2003

Erstgutachter:	Priv.-Doz. Dr. Volker Eyert
Zweitgutachter:	Prof. Dr. Thilo Kopp
Externer Gutachter:	Prof. Dr. Karlheinz Schwarz
Tag der Einreichung:	19. Dezember 2003
Tag der Prüfung:	23. Februar 2004

Vorwort

Das anhaltende Interesse der modernen Festkörperphysik an den Übergangsmetalloxiden beruht auf einer Vielzahl außergewöhnlicher Phänomene wie z. B. Metall-Isolator-Übergänge, strukturelle Umwandlungen und magnetische Ordnungen. Herr Schwingenschlögl stellt derartige Phänomene anhand einer Reihe sehr prominenter Beispiele eindrucksvoll dar. Beginnend mit den „klassischen“ Metall-Isolator-Übergängen der Vanadate VO_2 und V_2O_3, die bis heute nicht vollständig verstanden sind, spannt er mit seiner Untersuchung der Magnéli-Phasen erstmalig einen Bogen zwischen diesen beiden Verbindungen und legt so die Grundlage für ein umfassendes Verständnis der gesamten Materialklasse. In einem weiteren Teil der Arbeit erläutert Herr Schwingenschlögl anhand einer neuen Klasse von Kobaltaten das Zusammenspiel zwischen Frustrationseffekten und magnetischen Wechselwirkungen, welches Anlass zu komplizierter magnetischer Ordnung gibt. Die Kombination einer prägnanten Darstellung der theoretischen Grundlagen mit einem weiten Überblick über aktuelle Themen der Materialforschung macht die Dissertation von Herrn Schwingenschlögl zu einem sehr lesenswerten Werk.

Priv.-Doz. Dr. Volker Eyert, Augsburg, 03.03.2004

Abstract

In this thesis electronic structure calculations are applied to gain fundamental insight into physical mechanisms giving rise to specific material properties.
The metal-insulator transitions of the vanadium Magnéli phases V_nO_{2n-1} are studied by identifying the relevant electronic states and analyzing their response to structural modifications. This investigation is based on a systematic understanding of the various crystal structures, including those of VO_2 and V_2O_3. It is possible to study the relations between structural and electronic properties at the metal-insulator transitions of the latter oxides on a common basis. Similar crystal structures allow for the knowledge of the phase transitions in the vanadium oxides to be transferred to the titanium Magnéli phases.
Octahedral tiltings, as found in the ruthenates $ACu_3Ru_4O_{12}$, are the most important class of structural distortions affecting perovskites. Analyzing the relationship between tiltings and electronic features paves the way for a universal picture of octahedral tilting.
Furthermore, a family of quasi one-dimensional materials characterized by extraordinary magnetic properties is addressed. Motivated by band structure calculations for $Ca_3Co_2O_6$ the related compounds Ca_3CoRhO_6 and Ca_3FeRhO_6 are investigated, allowing for insight into the details of the magnetic coupling.

Kurzfassung

In der vorliegenden Dissertation werden Bandstrukturrechnungen eingesetzt, um Einblicke in physikalische Mechanismen zu erlangen, welche charakteristische Materialeigenschaften zur Folge haben.
Zur Untersuchung der Metall-Isolator-Übergänge der Vanadium Magnéli Phasen V_nO_{2n-1} werden zuerst die relevanten elektronischen Zustände identifiziert und anschließend deren Reaktionen auf strukturelle Modifikationen analysiert. Dieses Vorgehen basiert auf einem systematischen Verständnis der einzelnen Kristallstrukturen, einschließlich derjenigen von VO_2 sowie V_2O_3. Es ist möglich die Beziehungen zwischen den strukturellen und elektronischen Eigenschaften am Metall-Isolator-Übergang letztgenannter Oxide zu vergleichen. Ähnlichkeiten in der Kristallstruktur erlauben es, Erkenntnisse zu den Phasenübergängen der Vanadiumoxide auf die Titan Magnéli Phasen zu übertragen.
Oktaederverkippungen, wie sie in den Ruthenaten $ACu_3Ru_4O_{12}$ vorzufinden sind, bilden die wichtigste Klasse struktureller Verzerrungen der Perowskite. Eine detaillierte Analyse der Beziehungen zwischen solchen Verzerrungen und der elektronischen Struktur führt zu einem umfassenden Verständnis der zugrunde liegenden Mechanismen.
Weiterhin wird eine neue Familie quasi-eindimensionaler Materialien mit ungewöhnlichen magnetischen Eigenschaften betrachtet. Motiviert durch Bandstrukturrechnungen für den Vertreter $Ca_3Co_2O_6$ gilt das Interesse den Verbindungen Ca_3CoRhO_6 sowie Ca_3FeRhO_6. Eine vergleichende Untersuchung dieser Substanzen gestattet Einblicke in die Details der magnetischen Kopplung.

PACS: 71.20.-b, 71.27.+a, 71.30.+h, 72.15.Nj, 75.30.Et

Contents

Chapter 1

Introduction

For several decades, density functional theory has been established as a valuable 'ab initio' approach to describe condensed matter. Combined with the local density approximation the method allows accurate calculation of the ground state properties of materials. Starting just with the atomic arrangement of a compound it yields, for instance, the electronic, magnetic, or optical behaviour. From an application-oriented point of view density functional theory paves the way for tailoring materials and is an important tool in materials science. However, electronic structure calculations are likewise successfully used to gain fundamental insight into the physical mechanisms giving rise to the specific material properties. Aiming at an understanding of the latter, the calculated electronic structure may serve as a starting point to develop explanations in terms of suited models. In particular, it allows an investigation of the interplay of the characteristic structural features and the resulting physical properties. The 'first principles' studies discussed in this thesis proceed along that line. Expectedly, we can take advantage of evaluating the above interrelation in many fields of condensed matter physics and chemistry.

As is generally known, studying phase transitions has a long history in physics and is still an active field. A specific class is defined by the metal-insulator transitions in crystalline solids, which are not only of interest for basic reseach but also reveal special technological relevance. Strikingly, the changes of the conductivity often coincide with structural transformations. This is true particularly for temperature induced metal-insulator transitions found in transition metal oxides. The latter is a highly interesting material class as many compounds are characterized by narrow metal *d* bands in the vicinity of the Fermi level. As narrow *d* states are susceptible to electronic correlations, transition metal oxides are likely to show unusual electronic properties. In addition, the *d* states give rise to directed bonds thus paving way for strong electron-phonon coupling or degeneracy lifting processes as the Jahn–Teller effect and orbital ordering. Since various mechanisms have to be considered it is not surprising that controversial discussions concern the driving forces of the phase transitions.

In this context a large number of both experimental and theoretical investigations dealing with the metal-insulator transitions in the class of the vanadium oxides is reported in the literature. Special focus is on the prototypical compounds vanadium dioxide (VO_2) and sesquioxide (V_2O_3). However, much dispute remains concerning the origin of the phase

transitions in the two materials. While it is agreed that they trace back to a delicate interplay of electron-phonon coupling and electronic correlations, the relative importance of the mechanisms is a matter of ongoing discussions. To arrive at a deeper understanding of these transitions, an identification of the relevant electronic states and an investigation of their response to the changes of the crystal structure is desirable. This issue is addressed in the present thesis by analyzing the relations between the structural and the electronic properties in the broader class of the vanadium Magnéli phases. These materials form the homologous series V_nO_{2n-1} ($3 \leq n \leq 9$) and reveal crystal structures comprising typical dioxide-like and sesquioxide-like regions. They may be regarded as intermediate between the structures of the end members VO_2 ($n \to \infty$) and V_2O_3 ($n = 2$), allowing for insight into the crossover between the latter oxides. With one exception, the vanadium Magnéli phases exhibit metal-insulator transitions as functions of temperature. As the transitions are accompanied by structural transformations we relate the changes in the local atomic environments to the electronic behaviour. In the end this paves way to better understanding of the transport properties.
Because an interpretation of results obtained by electronic structure calculations requires some knowledge of the underlying concepts and methods, a short outline of the theoretical basis is presented in the second chapter. To prepare for an investigation of the vanadium Magnéli phases, chapter 3 summarizes results for both vanadium dioxide and sesquioxide. The crystal structures of the compounds and the corresponding electronic structures are investigated in detail. In chapter 4 the very first 'ab initio' study of the vanadium Magnéli phases is reported. The investigation is based on a new and unifying description of the crystal structures of all the compounds including VO_2 and V_2O_3. After developing this representation we address two specially suited members of the Magnéli series, V_4O_7 and V_6O_{11}, and discuss them in detail. Analyzing their local electronic properties permits us to approach the phase transitions of the series. Moreover, the findings shed new light on the role of particular electronic states for the metal-insulator transition of V_2O_3. To complete the investigation of the vanadium Magnéli phases we discuss the systematic aspects inherent in the homologous series.
In analogy to the vanadium compounds the related titanium oxides likewise give rise to a Magnéli series. The structural similarities of the homologous series allow complete transference of the knowledge developed for the phase transitions in the vanadium compounds to the titanium case. We use this relationship to investigate the metal-insulator transition of Ti_4O_7 in chapter 5. Here we focus on the origin and the effects of the distinct titanium pairing present in the insulating modification. The relations between the pairing and the phase transition still await a satisfactory explanation. In addition, a comparative analysis of the corundum based compounds V_2O_3 and Ti_2O_3 helps us to identify the origin of the frequently discussed quasi one-dimensional a_{1g} bands. These states play a major role in many theories addressing the metal-insulator transitions of V_2O_3.
In chapter 6 we turn to the perovskite-related materials $ACu_3Ru_4O_{12}$ (A=Na, Ca, Sr, La, Nd). Despite its striking simplicity the perovskite structure allows for a lot of crystallographic variations giving room for a wide class of compounds. Interest in these materials is motivated by exciting dielectric, magnetic, electrical, optical, or catalytic features with different technological applications. For optimal tailoring of the material properties much research on the perovskite-related compounds concentrates on the interrelation of the de-

viations from the ideal perovskite structure and the resulting physical properties. In this context octahedral tiltings as realized in the ruthenates $ACu_3Ru_4O_{12}$ represent the most important deviations. The present investigation deals with the relationship between the tiltings of the RuO_6 octahedra and the electronic features of the above ruthenates. This permits understanding of the specific bond lengths and the origin of the tilting in terms of chemical bonding. In doing so we can develop a universal picture of octahedral tilting in perovskite-related materials.

Low-dimensional systems are known to show most fascinating physical properties. In the last years it was possible to synthesize a new structural family of (quasi) one-dimensional materials characterized by extraordinary magnetic properties, which lack microscopic understanding. These magnetic chain compounds belong to the hexagonal perovskite oxides and are based on the general chemical formula $A_3A'BO_6$. A frequently discussed member of the class is $Ca_3Co_2O_6$, attracting interest due to possible partially disordered antiferromagnetism. Motivated by electronic structure calculations for this material, we analyze the related compounds Ca_3CoRhO_6 and Ca_3FeRhO_6 in chapter 7. Ca_3CoRhO_6 exhibits magnetic properties similar to those of $Ca_3Co_2O_6$ as it is characterized by ferromagnetic intrachain coupling complemented by antiferromagnetic interchain coupling. Because of its specific crystal structure the latter is affected by frustration effects. Unlike the peculiar magnetic ordering in the previous material we observe a rather simple antiferromagnetic-type behaviour in the case of Ca_3FeRhO_6. In order to investigate the microscopic origin of the magnetism in both compounds it is profitable to study the effects of exchanging the particular transition metal atoms. We will realize similarities in the electronic structures of $Ca_3Co_2O_6$, Ca_3CoRhO_6, and even Ca_3FeRhO_6, which allow for an interpretation of the magnetic coupling on a common basis.

Chapter 2

Density Functional Theory and its Application

The band structure calculations presented in the following chapters are performed using the density functional theory and the local density approximation. Hence we now aim at an introduction into the method and an outline of its specific features. Both is necessary since the correct interpretation of results obtained from electronic structure calculations requires some knowledge of the theoretical basis [1–8].

2.1 Electronic Hamiltonian

Let us begin by thinking of a solid as a composition of ions and valence electrons. The ions themselves are made up of nuclei and tightly bound core electrons. Each nucleus carries the charge $Z \cdot e$, where Z is the atomic number of the particular chemical element and e is the electronic charge. Subsequently we presume the core electrons not to contribute to the transport properties of the solid. Only the (mobile) valence electrons are responsible for effects such as the metallic conductivity. Of course, a strict separation between core and valence electrons is an idealization, but it allows us to describe the solid in terms of interacting ions and valence electrons. This argumentation directly yields a Hamiltonian $H = H_{\text{ion}} + H_{\text{el}} + H_{\text{ion-el}}$ consisting of three terms:
The ionic subsystem contains the kinetic energy of the ions and the potential energy due to ion-ion Coulomb interaction

$$H_{\text{ion}} = H_{\text{ion,kin}} + H_{\text{ion-ion}} = \sum_{\mu} \left(-\frac{\hbar^2}{2M_\mu} \nabla_\mu^2 + \frac{1}{2} \frac{e^2}{4\pi\epsilon_0} \sum_{\mu \neq \nu} \frac{Z_{\text{val},\mu} Z_{\text{val},\nu}}{|\mathbf{R}_\mu - \mathbf{R}_\nu|} \right) . \tag{2.1}$$

The electronic subsystem comprises the kinetic energy of the electrons and the potential energy due to electron-electron Coulomb interaction

$$H_{\text{el}} = H_{\text{el,kin}} + H_{\text{el-el}} = \sum_{i} \left(-\frac{\hbar^2}{2m} \nabla_i^2 + \frac{1}{2} \frac{e^2}{4\pi\epsilon_0} \sum_{i \neq j} \frac{1}{|\mathbf{r}_i - \mathbf{r}_j|} \right) . \tag{2.2}$$

Finally, the coupling of the ionic and the electronic subsystems is given by the potential energy due to ion-electron Coulomb interaction

$$H_{\text{ion-el}} = -\frac{e^2}{4\pi\epsilon_0}\sum_{\mu}\sum_{i}\frac{Z_{\text{val},\mu}}{|\mathbf{R}_\mu - \mathbf{r}_i|}\,. \tag{2.3}$$

Here M_μ and $\mathbf{R}_\mu$ as well as m and $\mathbf{r}_i$ represent the masses and positions of the ions and electrons, respectively. In addition, $Z_{\text{val},\mu}$ is the number of valence electrons provided by the ion at position $\mathbf{R}_\mu$.

The solution $|\Psi\rangle$ of the time-independent Schrödinger equation

$$H|\Psi\rangle = \mathcal{E}|\Psi\rangle \tag{2.4}$$

depends on the complete sets $\{\mathbf{R}_\mu\}$, $\{\mathbf{r}_i\}$ of ionic and electronic coordinates. A solution to the full problem containing both the ionic and electronic degrees of freedom is hardly achievable. However, a considerable simplification is provided by the Born–Oppenheimer approximation [9], which results in an effective decoupling of the ionic from the electronic problem. From a physical point of view this decoupling is a consequence of the fact that the ions are much heavier than the electrons. Hence the latter move much faster and for a given set of ionic sites the electrons adjust their positions almost instantaneously, i.e. they follow the ionic motion adiabatically. On the timescale of electronic motion the ions do not move at all but stay at fixed positions.
If the ions are fixed at positions $\{\mathbf{R}_\mu\}$, the Schrödinger equation for the electronic problem is given by

$$(H_{\text{el}} + H_{\text{ion-el}})\,\psi_\alpha(\{\mathbf{r}_i, \mathbf{R}_\mu\}) = E_\alpha(\{\mathbf{R}_\mu\})\psi_\alpha(\{\mathbf{r}_i, \mathbf{R}_\mu\})\,, \tag{2.5}$$

where ψ_α represents the eigenstates of the electronic Hamiltonian $H_{\text{el}+} = H_{\text{el}} + H_{\text{ion-el}}$. The eigenstates are orthogonal real functions, which are assumed to be normalized. The positions of the ions play the role of parameters in the electronic Hamiltonian $H_{\text{el}+}$, the wave functions ψ_α, and the eigenvalues E_α. Expanding the wave functions Ψ_γ belonging to the full Schrödinger equation in the complete set of solutions ψ_α of the electronic problem one obtains

$$\Psi_\gamma(\{\mathbf{r}_i, \mathbf{R}_\mu\}) = \sum_\alpha \phi_{\gamma\alpha}(\{\mathbf{R}_\mu\})\psi_\alpha(\{\mathbf{r}_i, \mathbf{R}_\mu\})\,. \tag{2.6}$$

Within the Born–Oppenheimer approximation this yields a Schrödinger equation for the ionic problem

$$(H_{\text{ion}} + E_\alpha(\{\mathbf{R}_\mu\}))\,\phi_{\gamma\alpha}(\{\mathbf{R}_\mu\}) = \mathcal{E}_\gamma\phi_{\gamma\alpha}(\{\mathbf{R}_\mu\})\,, \tag{2.7}$$

which is completely decoupled from the electronic degrees of freedom. The ionic Hamiltonian $H_{\text{ion}+} = H_{\text{ion}} + E_\alpha(\{\mathbf{R}_\mu\})$ contains the bare ionic contribution H_{ion} and in addition the total energy of the electronic system, which itself is a function of the ionic positions. The electronic Schrödinger equation (2.5) does not comprise the dynamics of the ionic motion, but the ions enter only via their fixed positions. Furthermore, the ionic Schrödinger equation (2.7) does not consider the specific electron distribution. Only the total energy of the electrons is included. Both these facts reflect the different time scales of the ionic and the electronic motion.

Summarizing, the Born–Oppenheimer approximation permits us to concentrate solely on the solution of the electronic problem. Because the ion-ion interaction just gives rise to a constant contribution to the electronic energies we treat it together with the electronic problem. Hence we have to discuss the Hamiltonian

$$H_0 = H_{\text{ion-ion}} + H_{\text{el}}(\{\mathbf{r}_i\}) + H_{\text{ion-el}}(\{\mathbf{r}_i\}) \tag{2.8}$$

in detail in the following sections. By merging the external (ionic) parts of this Hamiltonian, i.e. writing $H_{\text{ext}}(\{\mathbf{r}_i\}) = H_{\text{ion-ion}} + H_{\text{ion-el}}(\{\mathbf{r}_i\})$, we end up with the expression

$$\begin{aligned} H_0 &= H_{\text{el,kin}}(\{\mathbf{r}_i\}) + H_{\text{el-el}}(\{\mathbf{r}_i\}) + H_{\text{ext}}(\{\mathbf{r}_i\}) \\ &= \sum_i \left(-\frac{\hbar^2}{2m}\nabla_i^2\right) + \frac{1}{2}\frac{e^2}{4\pi\epsilon_0}\sum_{i\neq j}\frac{1}{|\mathbf{r}_i - \mathbf{r}_j|} \\ &\quad + \frac{1}{2}\frac{e^2}{4\pi\epsilon_0}\sum_{\mu\neq\nu}\frac{Z_{\text{val},\mu}Z_{\text{val},\nu}}{|\mathbf{R}_\mu - \mathbf{R}_\nu|} - \frac{e^2}{4\pi\epsilon_0}\sum_\mu\sum_i\frac{Z_{\text{val},\mu}}{|\mathbf{R}_\mu - \mathbf{r}_i|}\,. \end{aligned} \tag{2.9}$$

While the aim of the above was to introduce and simplify the Hamiltonian applied subsequently, we now turn to the solution of the electronic Schrödinger equation within density functional theory. Nowadays the latter approach plays an important role in understanding physical mechanisms of a multitude of phenomena in condensed matter physics. In many cases it allows an accurate calculation of the electronic structure, which is responsible for a large variety of material properties.
As the name suggests, the electron density ρ is the central variable in density functional theory. Therefore we define the electron density operator

$$\hat{\rho}(\mathbf{r}) = \sum_\sigma \hat{\rho}_\sigma(\mathbf{r}) = \sum_\sigma \hat{\psi}_\sigma^+(\mathbf{r})\hat{\psi}_\sigma(\mathbf{r})\,, \tag{2.10}$$

where σ denotes the spin index and $\hat{\psi}_\sigma^+$, $\hat{\psi}_\sigma$ are the field operators. Using a basis $\{|\chi_\alpha\rangle\}$ of single-particle states the latter are given by

$$\begin{aligned} \hat{\psi}_\sigma(\mathbf{r}) &= \sum_\alpha \langle\mathbf{r},\sigma|\chi_\alpha\rangle a_\alpha = \sum_\alpha \chi_{\alpha,\sigma}(\mathbf{r})a_\alpha \\ \hat{\psi}_\sigma^+(\mathbf{r}) &= \sum_\alpha \langle\mathbf{r},\sigma|\chi_\alpha\rangle^* a_\alpha^+ = \sum_\alpha \chi_{\alpha,\sigma}^*(\mathbf{r})a_\alpha^+\,, \end{aligned} \tag{2.11}$$

where the index α comprises all quantum numbers (expect for the spin σ) characterizing the single-particle state $|\chi_\alpha\rangle$. The fermionic creation and annihilation operators a_α^+, a_α create and annihilate electrons in the states $|\chi_\alpha\rangle$. Wave functions in real space representation are denoted $\chi_\alpha(\mathbf{r})$ and for their spin components we write $\chi_{\alpha,\sigma}(\mathbf{r})$. Bearing in mind equation (2.10) the electron density operator now takes the form

$$\hat{\rho}(\mathbf{r}) = \sum_\sigma\sum_{\alpha,\alpha'} \chi_{\alpha,\sigma}^*(\mathbf{r})\chi_{\alpha',\sigma}(\mathbf{r})a_\alpha^+ a_{\alpha'}\,. \tag{2.12}$$

The electron density $\rho(\mathbf{r})$ is calculated as

$$\rho(\mathbf{r}) = \sum_\sigma \rho_\sigma(\mathbf{r}) = \sum_\sigma \langle\Psi|\hat{\rho}_\sigma(\mathbf{r})|\Psi\rangle\,, \tag{2.13}$$

where $|\Psi\rangle$ denotes the electronic many-body wave function. Since the latter reduces to a single Slater determinant for non-interacting particles we may write the electron density in terms of single-particle states

$$\rho(\mathbf{r}) = \sum_{\alpha,\sigma}^{\text{occ}} |\chi_{\alpha,\sigma}(\mathbf{r})|^2 \,. \tag{2.14}$$

The second sum comprises all occupied states. Integrating $\rho(\mathbf{r})$ over the full space yields the total number N of electrons in the system

$$\int d^3\mathbf{r}\; \rho(\mathbf{r}) = N \,. \tag{2.15}$$

As a consequence of the field quantization we can express the Hamiltonian H_0 in terms of the field operators $\hat{\psi}_\sigma^+$, $\hat{\psi}_\sigma$ and the electron density operator $\hat{\rho}$

$$H_{\text{el,kin}} = \sum_\sigma \int d^3\mathbf{r}\; \hat{\psi}_\sigma^+(\mathbf{r}) \left(-\frac{\hbar^2}{2m}\nabla^2\right) \hat{\psi}_\sigma(\mathbf{r}) \tag{2.16}$$

$$H_{\text{el-el}} = \frac{1}{2}\frac{e^2}{4\pi\epsilon_0} \sum_{\sigma,\sigma'} \int d^3\mathbf{r} \int d^3\mathbf{r}'\; \hat{\psi}_\sigma^+(\mathbf{r})\hat{\psi}_{\sigma'}^+(\mathbf{r}') \frac{1}{|\mathbf{r}-\mathbf{r}'|} \hat{\psi}_{\sigma'}(\mathbf{r}')\hat{\psi}_\sigma(\mathbf{r}) \tag{2.17}$$

$$\begin{aligned} H_{\text{ext}} &= \int d^3\mathbf{r} \left(\frac{1}{2N}\frac{e^2}{4\pi\epsilon_0} \sum_{\mu\neq\nu} \frac{Z_{\text{val},\mu} Z_{\text{val},\nu}}{|\mathbf{R}_\mu - \mathbf{R}_\nu|} - \frac{e^2}{4\pi\epsilon_0} \sum_\mu \frac{Z_{\text{val},\mu}}{|\mathbf{R}_\mu - \mathbf{r}|} \right) \hat{\rho}(\mathbf{r}) \\ &= \int d^3\mathbf{r}\; v_{\text{ext}}(\mathbf{r}) \hat{\rho}(\mathbf{r}) \,. \end{aligned} \tag{2.18}$$

Here the external potential $v_{\text{ext}}(\mathbf{r})$ was introduced to simplify the notation in the following sections.

2.2 Hohenberg–Kohn Theorems and Implications

Density functional theory recasts the problem of calculating the electronic (many-body) wave function of the ground state in terms of the electron density distribution $\rho(\mathbf{r})$ and a universal energy functional $E_{\text{xc}}[\rho(\mathbf{r})]$ of this density – the exchange-correlation functional. The task of solving the many electron Schrödinger equation is replaced by the problem of finding accurate approximations to the energy functional and solving effective single electron equations. Regarding the electronic ground state, density functional theory formally can be interpreted as further development of the electron density approach beyond Thomas–Fermi theory and as an alternative to the Hartree–Fock method, which neglects electronic correlations. In this section some details of the density functional approach are discussed. To this end we start with two basic theorems, which trace back to P. Hohenberg and W. Kohn [10]. In their original work these authors deal with a spin independent potential and assume the ground state to be non-degenerate. M. Levy [11] later overcame the restriction to non-degenerate ground states, which is important from the formal point of view. However, for clarity we consider in the following the non-degenerate case.

Theorem 1:
The ground state electron density $\rho_0(\mathbf{r})$ determines the external potential $v_{\text{ext}}(\mathbf{r})$.

Theorem 2:
The total energy functional $E[\rho(\mathbf{r})]$ has a minimum at the correct ground state electron density $\rho_0(\mathbf{r})$

$$E[\rho_0(\mathbf{r})] < E[\rho(\mathbf{r})]\,. \tag{2.19}$$

In order to prove theorem 1, we first show that two external potentials $v_{\text{ext}}(\mathbf{r})$ and $v'_{\text{ext}}(\mathbf{r})$, which differ by more than a trivial constant, necessarily yield different ground states $|\Psi_0\rangle$ and $|\Psi'_0\rangle$. The Schrödinger equations for the latter are given by

$$[H_{\text{el,kin}} + H_{\text{el-el}} + H_{\text{ext}}]|\Psi_0\rangle = \mathcal{E}_0|\Psi_0\rangle\,, \quad [H_{\text{el,kin}} + H_{\text{el-el}} + H'_{\text{ext}}]|\Psi'_0\rangle = \mathcal{E}'_0|\Psi'_0\rangle\,, \tag{2.20}$$

where $\mathcal{E}_0$ and $\mathcal{E}'_0$ are the ground state energies. If we suppose $|\Psi_0\rangle = |\Psi'_0\rangle$, we obtain

$$[H_{\text{ext}} - H'_{\text{ext}}]|\Psi_0\rangle = [\mathcal{E}_0 - \mathcal{E}'_0]|\Psi_0\rangle\,. \tag{2.21}$$

As a consequence, the external potentials differ at most by a constant, which contradicts our assumption. We have thus shown that if $v_{\text{ext}}(\mathbf{r}) \neq v'_{\text{ext}}(\mathbf{r})$ then $|\Psi_0\rangle \neq |\Psi'_0\rangle$. However, we still have to prove that the latter yields $\rho_0(\mathbf{r}) \neq \rho'_0(\mathbf{r})$ for the associated ground state electron densities $\rho_0(\mathbf{r})$ and $\rho'_0(\mathbf{r})$. Therefore we use the variational principle of quantum mechanics

$$\begin{aligned}\mathcal{E}_0 &= \langle\Psi_0|H_{\text{el,kin}} + H_{\text{el-el}} + H_{\text{ext}}|\Psi_0\rangle \\ &< \langle\Psi'_0|H_{\text{el,kin}} + H_{\text{el-el}} + H_{\text{ext}}|\Psi'_0\rangle = \mathcal{E}'_0 + \int d^3\mathbf{r}\,[v_{\text{ext}}(\mathbf{r}) - v'_{\text{ext}}(\mathbf{r})]\rho'_0(\mathbf{r})\,.\end{aligned} \tag{2.22}$$

Analogous arguments lead to

$$\mathcal{E}'_0 < \mathcal{E}_0 + \int d^3\mathbf{r}\,[v'_{\text{ext}}(\mathbf{r}) - v_{\text{ext}}(\mathbf{r})]\rho_0(\mathbf{r})\,. \tag{2.23}$$

Assuming $\rho_0(\mathbf{r}) = \rho'_0(\mathbf{r})$ and adding equations (2.22) and (2.23) yields $\mathcal{E}_0 + \mathcal{E}'_0 < \mathcal{E}'_0 + \mathcal{E}_0$, which is a contradiction. We have thus established that different (non-degenerate) ground states lead to different ground state densities. Identical densities consequently stem from identical external potentials, which ends the proof of theorem 1.
Once the external potential $v_{\text{ext}}(\mathbf{r})$ is specified, the Hamiltonian H_0 is entirely determined. The eigenstates and the ground state density $\rho_0(\mathbf{r})$ consequently are known (in principle). Due to theorem 1 the mapping from the external potential to the ground state density is invertible. Given the ground state density of any N-electron system, the Hamiltonian of the system is uniquely determined and we know (in principle) all the eigenstates and the expectation value of any operator. In particular, the ground state energy $\mathcal{E}_0$ is a unique functional of $\rho_0(\mathbf{r})$. We define the total energy functional as

$$E[\rho(\mathbf{r})] = \langle\Psi|H_{\text{el,kin}} + H_{\text{el-el}} + v_{\text{ext}}(\mathbf{r})|\Psi\rangle\,, \tag{2.24}$$

where $v_{\text{ext}}(\mathbf{r})$ is the specific external potential of a system with ground state density $\rho_0(\mathbf{r})$ and ground state energy $\mathcal{E}_0$. Moreover, $|\Psi\rangle$ corresponds to the ground state density $\rho(\mathbf{r})$ of some N-electron system. If $\rho(\mathbf{r}) = \rho_0(\mathbf{r})$, the total energy functional $E[\rho(\mathbf{r})]$ takes the value $\mathcal{E}_0$. Since the ground state energy is uniquely determined by $\rho_0(\mathbf{r})$, the variational principle establishes for $\rho(\mathbf{r}) \neq \rho_0(\mathbf{r})$

$$\mathcal{E}_0 = E[\rho_0(\mathbf{r})] < E[\rho(\mathbf{r})] \,. \tag{2.25}$$

The ground state energy therefore can be obtained by varying the density to minimize the energy, provided we know the total energy functional, or at least a good approximation. Because the Hamiltonians representing the kinetic energy of the electrons (2.16) and the electron-electron interaction (2.17) do not depend on the external potential,

$$F[\rho(\mathbf{r})] = \langle\Psi|H_{\text{el,kin}} + H_{\text{el}-\text{el}}|\Psi\rangle \tag{2.26}$$

is a universal functional of the electron density $\rho(\mathbf{r})$. Bearing in mind equation (2.18) we find

$$\langle\Psi|H_{\text{ext}}|\Psi\rangle = \int d^3\mathbf{r}\; v_{\text{ext}}(\mathbf{r})\rho(\mathbf{r}) \tag{2.27}$$

and therefore the total energy functional can be written as

$$E[\rho(\mathbf{r})] = F[\rho(\mathbf{r})] + \int d^3\mathbf{r}\; v_{\text{ext}}(\mathbf{r})\rho(\mathbf{r}) \,. \tag{2.28}$$

In order to have a closer look at the universal functional $F[\rho(\mathbf{r})]$ given by equation (2.26) it is useful to introduce the kinetic energy functional

$$T[\rho(\mathbf{r})] = \langle\Psi|H_{\text{el,kin}}|\Psi\rangle \,. \tag{2.29}$$

Separating the direct (Hartree) term from the electron-electron Coulomb interaction one finds

$$\begin{aligned} E[\rho(\mathbf{r})] &= T[\rho(\mathbf{r})] + \frac{1}{2}\frac{e^2}{4\pi\epsilon_0}\sum_{\sigma,\sigma'}\int d^3\mathbf{r}\int d^3\mathbf{r}'\,\frac{\rho_\sigma(\mathbf{r})\rho_{\sigma'}(\mathbf{r}')}{|\mathbf{r}-\mathbf{r}'|} \\ &\quad + \int d^3\mathbf{r}\; v_{\text{ext}}(\mathbf{r})\rho(\mathbf{r}) + W_{\text{xc}}[\rho(\mathbf{r})] \,, \end{aligned} \tag{2.30}$$

where the functional $W_{\text{xc}}[\rho(\mathbf{r})]$ comprises all the energy contributions due to the Hamiltonian $H_{\text{el}-\text{el}}$ (2.17) not covered by the direct term. It contains the exchange term of the Hartree–Fock theory [12]

$$W_{\text{x}}[\rho(\mathbf{r})] = -\frac{1}{2}\frac{e^2}{4\pi\epsilon_0}\sum_{\sigma,\sigma'}\int d^3\mathbf{r}\int d^3\mathbf{r}'\,\frac{|\rho_{\sigma,\sigma'}(\mathbf{r},\mathbf{r}')|^2}{|\mathbf{r}-\mathbf{r}'|} \tag{2.31}$$

and all the correlation contributions $W_{\text{c}}[\rho(\mathbf{r})]$ which emerge when one goes beyond the Hartree–Fock approximation. In equation (2.31) we applied the spin dependent density matrix $\rho_{\sigma,\sigma'}(\mathbf{r},\mathbf{r}')$ defined by

$$\rho_{\sigma,\sigma'}(\mathbf{r},\mathbf{r}') = \langle\Psi|\hat{\rho}_{\sigma,\sigma'}(\mathbf{r},\mathbf{r}')|\Psi\rangle = \langle\Psi|\hat{\psi}^+_{\sigma'}(\mathbf{r}')\hat{\psi}_\sigma(\mathbf{r})|\Psi\rangle \,. \tag{2.32}$$

While theorem 1 yields the correctness of the electron density approach, theorem 2 gives rise to a practical scheme, which eventually allows calculation of the ground state properties from the electron density.

As a consequence of the discussed minimum of the total energy functional at the correct ground state density W. Kohn and L. J. Sham [13] were able to apply a variational principle and subsequently derived effective single-particle equations as follows.
Starting from equation (2.30) for the total energy functional and taking into account the particle conservation (2.15) by introducing a Lagrange multiplier μ (chemical potential) the variational principle is formulated as

$$\frac{\delta}{\delta\rho(\mathbf{r})}\left(E[\rho(\mathbf{r})] - \mu \int d^3\mathbf{r}\, \rho(\mathbf{r})\right) = 0\,, \tag{2.33}$$

from which we arrive at the Euler–Lagrange equation

$$\frac{\delta T[\rho(\mathbf{r})]}{\delta\rho(\mathbf{r})} + v_{\mathrm{H}}(\mathbf{r}) + v_{\mathrm{ext}}(\mathbf{r}) + \frac{\delta W_{\mathrm{xc}}[\rho(\mathbf{r})]}{\delta\rho(\mathbf{r})} = \mu \tag{2.34}$$

with the Hartree potential $v_{\mathrm{H}}(\mathbf{r})$ defined by

$$v_{\mathrm{H}}(\mathbf{r}) = \frac{e^2}{4\pi\epsilon_0} \int d^3\mathbf{r}'\, \frac{\rho(\mathbf{r}')}{|\mathbf{r} - \mathbf{r}'|}\,. \tag{2.35}$$

At this point W. Kohn and L. J. Sham use the trick of treating a fictitious system of non-interacting particles, which is assumed to have the same ground state density as the real system. The fictitious system is connected to some effective potential $v_{\mathrm{eff}}(\mathbf{r})$, the form of which we will determine in the following. In treating a system without electron-electron interaction we can deal with single-particle states $\{|\chi_\alpha\rangle\}$ and single-particle energies ϵ_α, connected by the Schrödinger equation

$$\left(-\frac{\hbar^2}{2m}\nabla^2 + v_{\mathrm{eff}}(\mathbf{r})\right)\chi_{\alpha,\sigma}(\mathbf{r}) = \epsilon_\alpha \chi_{\alpha,\sigma}(\mathbf{r})\,. \tag{2.36}$$

The total energy functional of the non-interacting system reduces to

$$\tilde{E}[\rho(\mathbf{r})] = \tilde{T}[\rho(\mathbf{r})] + \int d^3\mathbf{r}\; v_{\mathrm{eff}}(\mathbf{r})\rho(\mathbf{r}) \tag{2.37}$$

and due to the introduction of the single-particle wave functions the kinetic energy functional $\tilde{T}[\rho(\mathbf{r})]$ of the non-interacting system can be evaluated. Bearing in mind equations (2.11) and (2.16) one ends with the expression

$$\tilde{T}[\rho(\mathbf{r})] = \sum_{\alpha,\sigma}^{\mathrm{occ}} \int d^3\mathbf{r}\; \chi^*_{\alpha,\sigma}(\mathbf{r}) \left(-\frac{\hbar^2}{2m}\nabla^2\right) \chi_{\alpha,\sigma}(\mathbf{r})\,. \tag{2.38}$$

It is important to note that a potential $v_{\mathrm{eff}}(\mathbf{r})$ has to exist which gives rise to appropriate single-particle states. However, no further problems arise from this assumption [5].

We write the unknown functionals $T[\rho(\mathbf{r})]$ and $W_{\text{xc}}[\rho(\mathbf{r})]$ from equation (2.30) in terms of $\tilde{T}[\rho(\mathbf{r})]$ by using the likewise unknown exchange-correlation functional $E_{\text{xc}}[\rho(\mathbf{r})]$

$$T[\rho(\mathbf{r})] + W_{\text{xc}}[\rho(\mathbf{r})] = \tilde{T}[\rho(\mathbf{r})] + E_{\text{xc}}[\rho(\mathbf{r})]\,, \tag{2.39}$$

leading to a slightly modified total energy functional

$$\begin{aligned} E[\rho(\mathbf{r})] &= \sum_{\sigma}\sum_{\alpha}^{\text{occ}} \int d^3\mathbf{r}\; \chi^*_{\alpha,\sigma}(\mathbf{r}) \left(-\frac{\hbar^2}{2m}\nabla^2\right) \chi_{\alpha,\sigma}(\mathbf{r}) \\ &\quad + \frac{1}{2}\frac{e^2}{4\pi\epsilon_0} \sum_{\sigma,\sigma'} \int d^3\mathbf{r} \int d^3\mathbf{r}'\; \frac{\rho_\sigma(\mathbf{r})\rho_{\sigma'}(\mathbf{r}')}{|\mathbf{r}-\mathbf{r}'|} \\ &\quad + \int d^3\mathbf{r}\; v_{\text{ext}}(\mathbf{r})\rho(\mathbf{r}) + E_{\text{xc}}[\rho(\mathbf{r})]\,. \end{aligned} \tag{2.40}$$

To minimize the total energy functional by varying $\rho(\mathbf{r})$ we insert the density of the non-interacting problem, see equation (2.14). Recall that the ground state densities of the real and the fictitious system coincide by definition. Using the density of the non-interacting system in the variation of course does **not** mean that we assume the wave function of the interacting problem to be given by a Slater determinant. Achieving particle conservation via the normalization of the single-particle wave functions using the Lagrange multipliers ϵ_α we obtain the Euler–Lagrange equations

$$\begin{aligned} &\frac{\delta}{\delta\chi^*_{\alpha,\sigma}(\mathbf{r})}\left(E[\rho(\mathbf{r})] - \sum_{\alpha}\epsilon_\alpha \int d^3\mathbf{r}\; |\chi_{\alpha,\sigma}(\mathbf{r})|^2\right) \\ &= \left(-\frac{\hbar^2}{2m}\nabla^2 + v_{\text{H}}(\mathbf{r}) + v_{\text{ext}}(\mathbf{r}) + v_{\text{xc}}(\mathbf{r}) - \epsilon_\alpha\right)\chi_{\alpha,\sigma}(\mathbf{r}) = 0\,, \end{aligned} \tag{2.41}$$

where the exchange-correlation potential is deduced from the exchange correlation functional

$$v_{\text{xc}}(\mathbf{r}) = \frac{\delta E_{\text{xc}}[\rho(\mathbf{r})]}{\delta\rho(\mathbf{r})}\,. \tag{2.42}$$

Comparing equations (2.36) and (2.41) establishes the effectice potential $v_{\text{eff}}(\mathbf{r})$ defining the non-interacting system

$$v_{\text{eff}}(\mathbf{r}) = v_{\text{H}}(\mathbf{r}) + v_{\text{ext}}(\mathbf{r}) + v_{\text{xc}}(\mathbf{r})\,. \tag{2.43}$$

The Euler–Lagrange equations (2.41) are called Kohn–Sham equations. Formally, they are identical to the single-particle Schrödinger equation (2.36) with $v_{\text{eff}}(\mathbf{r})$ as stated in equation (2.43). Therefore solving the Kohn–Sham equations for a given effective potential is rather simple, but because the potential still contains the electronic charge density, a selfconsistent procedure is necessary. The solution of the Kohn–Sham equations yields single-particle wave functions and therefore a new density distribution, which again modifies the effective potential.
Applying a fictitious system of non-interacting particles allows us to map a complicated many-body problem onto an effective single-particle problem. However, the single-particle

states growing out of the Kohn–Sham equations lack any physical meaning. The same is true for the energies ϵ_α, which are nothing else than Lagrange multipliers. Only the energy of the highest occupied orbital is exceptional, since it is equal to the exact ionization potential [7, 14, 15]. Interpreting the other single-particle energies in terms of excitation energies is formally not justified but in practice often found to be correct.
The Hohenberg–Kohn theorems as denoted previously apply to the ground state of a system, which is supposed to not be spinpolarized. If the electron density is separated into one component for the spin up states and one for the spin down states, the Hohenberg–Kohn theorems have to be extended. This generalization was first analyzed by U. von Barth and L. Hedin [16] and for relativistic systems by A. K. Rajagopal and J. Callaway [17]. Spinpolarization might be a consequence of a symmetry breaking contribution to the external potential – for instance an external magnetic field – or it arises from the electron-electron interaction and therefore has to be traced back to the exchange-correlation potential. In each case we have to deal with real two component spinor wave functions and hence with 2×2 matrices in the Hamiltonian. Proceeding similarly to the case without spinpolarization modified Kohn–Sham equations are deduced. Instead of separated (and identical) equations for both spin directions, as in the spin degenerate case, we find coupled equations for the spin up and spin down components of the single-particle wave functions

$$\sum_{\sigma'} \left(-\delta_{\sigma,\sigma'} \frac{\hbar^2}{2m} \nabla^2 + v_{\text{eff}}^{\sigma,\sigma'}(\mathbf{r}) - \delta_{\sigma,\sigma'} \epsilon_{\alpha,\sigma} \right) \chi_{\alpha,\sigma'}(\mathbf{r}) = 0 \,. \tag{2.44}$$

Here we have defined a generalized effective potential matrix

$$v_{\text{eff}}^{\sigma,\sigma'}(\mathbf{r}) = \delta_{\sigma,\sigma'} v_{\text{H}}(\mathbf{r}) + v_{\text{ext}}^{\sigma,\sigma'}(\mathbf{r}) + v_{\text{xc}}^{\sigma,\sigma'}(\mathbf{r}) \,, \tag{2.45}$$

from which effects of spinpolarization can emerge due to possible spin dependence of the external and the exchange-correlation potential. Note that the Lagrange multipliers $\epsilon_{\alpha,\sigma}$ from the extended Kohn–Sham equations (2.44) explicitly depend on the spin index σ.

2.3 Local Density Approximation

In the preceding sections we introduced the density functional formalism in some detail. Admittedly, at this point we are not able to apply the scheme to realistic systems. This is due to the fact that the exchange-correlation functional entering the total energy (2.40) is completely unknown. However, a useful approximation to the functional is provided by the local density approximation (LDA), where the homogeneous interacting electron gas is used to model the unknown energy contributions. Therefore the exchange-correlation functional is expressed in terms of an integral over the local exchange correlation energy density $\epsilon_{\text{xc}}(\rho(\mathbf{r}))$ weighted with the local electron density $\rho(\mathbf{r})$

$$E_{\text{xc}}[\rho(\mathbf{r})] = \int d^3\mathbf{r} \; \rho(\mathbf{r}) \epsilon_{\text{xc}}(\rho(\mathbf{r})) \,. \tag{2.46}$$

The local electron density $\rho(\mathbf{r})$ is evaluated from the respective expression for the homogeneous interacting electron gas where the electron density is constant in space. By means

of equation (2.46) the main principle of the local density approach becomes obvious. The inhomogeneous electron system is divided into small regions located at positions $\mathbf{r}$, each containing a homogeneous interacting electron gas with some average density $\rho(\mathbf{r})$. One is now able to calculate the energy density and the local potential for every single region. Combining the results for all regions yields a description of the originally inhomogeneous system as a superposition of locally homogeneous electron gases. Apparently, the local density approximation is exact in the limit of a constant charge density. Furthermore, we expect an appropriate description of systems with slowly varying densities. But actually experience shows that the local density approximation works well in even more cases, for example for solids, where we are confronted with substantial deviations from a homogeneous electron distribution. Compared to semiconductors or insulators, metals are much better accounted for by the LDA due to a high mobility of the charge carriers.
For a spinpolarized ground state we have to handle spin up and spin down electron densities $\rho_\uparrow(\mathbf{r})$, $\rho_\downarrow(\mathbf{r})$. Instead of applying these quantities we use the total electron density $\rho(\mathbf{r})$ and the spin polarization $\zeta(\mathbf{r})$ to characterize the system

$$\rho(\mathbf{r}) = \rho_\uparrow(\mathbf{r}) + \rho_\downarrow(\mathbf{r}) \,, \quad \zeta(\mathbf{r}) = \frac{\rho_\uparrow(\mathbf{r}) - \rho_\downarrow(\mathbf{r})}{\rho(\mathbf{r})} \,. \tag{2.47}$$

In the spinpolarized case equation (2.46) takes the generalized form

$$E_{\mathrm{xc}}[\rho(\mathbf{r}), \zeta(\mathbf{r})] = \int d^3\mathbf{r} \; \rho(\mathbf{r}) \epsilon_{\mathrm{xc}}(\rho(\mathbf{r}), \zeta(\mathbf{r})) \,. \tag{2.48}$$

The connection between the spindependent exchange-correlation potential $v_{\mathrm{xc},\sigma}(\rho(\mathbf{r}), \zeta(\mathbf{r}))$ and the exchange-correlation functional is given by

$$v_{\mathrm{xc},\sigma}(\rho(\mathbf{r}), \zeta(\mathbf{r})) = \frac{\delta E_{\mathrm{xc}}[\rho(\mathbf{r}), \zeta(\mathbf{r})]}{\delta \rho_\sigma(\mathbf{r})} \,. \tag{2.49}$$

Finally, the energy density $\epsilon_{\mathrm{xc}}(\rho(\mathbf{r}), \zeta(\mathbf{r}))$ has to be expressed in terms of the charge density and the spin polarization [16,18]. For this purpose let us first neglect all correlation effects and start from the Hartree–Fock approximation for a spinpolarized electron gas. Applying atomic (Rydberg) units this results in the exchange energy density

$$\epsilon_{\mathrm{x}}(\rho(\mathbf{r}), \zeta(\mathbf{r})) = -6 \left(\frac{3}{4\pi} \right)^{\frac{1}{3}} \rho(\mathbf{r})^{-1} \left(\rho_\uparrow(\mathbf{r})^{\frac{4}{3}} + \rho_\downarrow(\mathbf{r})^{\frac{4}{3}} \right) \,. \tag{2.50}$$

Correlation effects give rise to the correlation energy density $\epsilon_{\mathrm{c}}(\rho(\mathbf{r}), \zeta(\mathbf{r}))$, which must be added to the exchange contribution

$$\epsilon_{\mathrm{xc}}(\rho(\mathbf{r}), \zeta(\mathbf{r})) = \epsilon_{\mathrm{x}}(\rho(\mathbf{r}), \zeta(\mathbf{r})) + \epsilon_{\mathrm{c}}(\rho(\mathbf{r}), \zeta(\mathbf{r})) \,. \tag{2.51}$$

At this point one is confronted with the problem that the correlation energy density of the homogeneous but interacting electron gas is not known exactly. Fortunately, numerically derived results exist, which are of high accuracy and thus provide useful approximations. Nowadays one may choose from a variety of parametrizations for the energy density, resulting from the pertubation theory or from Monte Carlo simulations (for details see for example [19]). The schemes introduced by U. von Barth and L. Hedin [16], S. H. Vosko, L. Wilk, and M. Nusair [20], or J. P. Perdew and A. Zunger [21] are frequently used. The former authors express the polarization dependence of the exchange and correlation energy densities of the homogeneous electron gas between the paramagnetic (P; $\zeta(\mathbf{r}) = 0$) and the saturated ferromagnetic (F; $\zeta(\mathbf{r}) = 1$) state in terms of a so-called spin interpolation formula

$$\begin{aligned} \epsilon_{\mathrm{x}}(\rho(\mathbf{r}),\zeta(\mathbf{r})) &= \epsilon_{\mathrm{x}}^{P}(\rho(\mathbf{r})) + \left(\epsilon_{\mathrm{x}}^{F}(\rho(\mathbf{r})) - \epsilon_{\mathrm{x}}^{P}(\rho(\mathbf{r}))\right) f(\zeta(\mathbf{r})) & (2.52) \\ \epsilon_{\mathrm{c}}(\rho(\mathbf{r}),\zeta(\mathbf{r})) &= \epsilon_{\mathrm{c}}^{P}(\rho(\mathbf{r})) + \left(\epsilon_{\mathrm{c}}^{F}(\rho(\mathbf{r})) - \epsilon_{\mathrm{c}}^{P}(\rho(\mathbf{r}))\right) f(\zeta(\mathbf{r}))\,. & (2.53) \end{aligned}$$

Here the spin interpolation function $f(\zeta(\mathbf{r}))$ is defined as

$$f(\zeta(\mathbf{r})) = \frac{1}{2\left(2^{\frac{1}{3}}-1\right)}\left(\left(\frac{2\rho_{\uparrow}(\mathbf{r})}{\rho(\mathbf{r})}\right)^{\frac{4}{3}} + \left(\frac{2\rho_{\downarrow}(\mathbf{r})}{\rho(\mathbf{r})}\right)^{\frac{4}{3}} - 2\right) \quad (2.54)$$

and vanishes if the spin up density is equal to the spin down density. Both in the paramagnetic and in the ferromagnetic case equation (2.50) can be simplified, resulting in

$$\epsilon_{\mathrm{x}}^{P}(\rho(\mathbf{r})) = -6\left(\frac{3\rho(\mathbf{r})}{8\pi}\right)^{\frac{1}{3}}, \quad \epsilon_{\mathrm{x}}^{F}(\rho(\mathbf{r})) = 2^{\frac{1}{3}}\epsilon_{\mathrm{x}}^{P}(\rho(\mathbf{r}))\,. \quad (2.55)$$

Using these relations only the problem of evaluating the density dependences of $\epsilon_{\mathrm{c}}^{P}(\rho(\mathbf{r}))$ and $\epsilon_{\mathrm{c}}^{F}(\rho(\mathbf{r}))$ remains open. In this context U. von Barth and L. Hedin proposed modeling the correlation energy density as

$$\epsilon_{\mathrm{c}}^{P}(r_S(\mathbf{r})) = -c_P \cdot F\left(\frac{r_S(\mathbf{r})}{r_P}\right), \quad \epsilon_{\mathrm{c}}^{F}(r_S(\mathbf{r})) = -c_F \cdot F\left(\frac{r_S(\mathbf{r})}{r_F}\right) \quad (2.56)$$

with the constants $c_P = 0.0504$, $r_P = 30$, $c_F = 0.0254$, and $r_F = 75$. The function $F(x)$ has the form

$$F(x) = (1+x^3)\ln\left(1+\frac{1}{x}\right) + \frac{x}{2} - x^2 - \frac{1}{3} \quad (2.57)$$

and the dimensionless density parameter $r_S(\mathbf{r})$ in units of the Bohr radius a_B is connected to the electron density by the relation

$$\frac{4\pi}{3} r_S(\mathbf{r})^3 a_B^3 = \frac{1}{\rho(\mathbf{r})}\,. \quad (2.58)$$

Applying the above preparations one calculates the exchange-correlation potential from equation (2.49) yielding [18]

$$\begin{aligned} v_{\mathrm{xc},\sigma}(\rho,\zeta) = {} & \frac{4}{3}\left(\epsilon_{\mathrm{x}}^{P}(r_S) + \frac{1}{2^{\frac{1}{3}}-1}\left(\epsilon_{\mathrm{c}}^{F}(r_S) - \epsilon_{\mathrm{c}}^{P}(r_S)\right)\right)\left(\frac{2\rho_\sigma}{\rho}\right)^{\frac{1}{3}} \\ & + \left(\mu_{\mathrm{c}}^{F}(r_S) - \mu_{\mathrm{c}}^{P}(r_S) - \frac{4}{3}\left(\epsilon_{\mathrm{c}}^{F}(r_S) - \epsilon_{\mathrm{c}}^{P}(r_S)\right)\right) f(\zeta) \\ & + \mu_{\mathrm{c}}^{P}(r_S) - \frac{4}{3}\frac{1}{2^{\frac{1}{3}}-1}\left(\epsilon_{\mathrm{c}}^{F}(r_S) - \epsilon_{\mathrm{c}}^{P}(r_S)\right) \end{aligned} \quad (2.59)$$

with the abbreviations

$$\mu_{\mathrm{c}}^{P}(r_S) = -c_P \cdot \ln\left(1 + \frac{r_P}{r_S}\right) , \quad \mu_{\mathrm{c}}^{F}(r_S) = -c_F \cdot \ln\left(1 + \frac{r_F}{r_S}\right) . \tag{2.60}$$

As already mentioned, several parametrizations to describe the exchange correlation density exist. But naturally none of them can overcome the local density approximation itself (and in particular the associated errors). To deal with this problem several concepts have been proposed. A first step of introducing dependences on spacial density variations is to allow the exchange-correlation potential to depend on the gradient of the electron density. Thereby one arrives at the so-called non-local density approximation or generalized gradient approximation, see J. P. Perdew in [4]. Although the generalized gradient approach leads to an improvement with respect to structural properties, it does not solve the band gap problem associated with the LDA, see below.

The interpretation of the Kohn–Sham energies ϵ_α given by equation (2.41) principally is difficult, which has consequences for the calculation of the band gap of a semiconductor or an insulator within the density functional approach. To tackle this point in more detail we have to relate the band gap Δ to ground-state properties, which the density functional formalism subsequently may deal with. Hence let $E(N)$ be the ground-state energy of a N-electron system, where N is continuous in general. In terms of the ionization potential I and the electron affinity energy A the band gap is defined as

$$\Delta = I - A = [E(N-1) - E(N)] - [E(N) - E(N+1)] . \tag{2.61}$$

Following articles of J. P. Perdew and M. Levy [22] as well as L. J. Sham and M. Schlüter [23] the band gap is connected to the eigenvalue gap $\Delta\epsilon$ of the N-electron density functional calculation by the relation

$$\Delta = \Delta\epsilon + \Delta_{\mathrm{xc}} = \epsilon_{N+1}(N) - \epsilon_N(N) + \Delta_{\mathrm{xc}} . \tag{2.62}$$

The equation states that the true band gap Δ might differ from the exact eigenvalue gap $\Delta\epsilon$ because of possible discontinuities in the exchange correlation potential $v_{\mathrm{xc}}(\mathbf{r}, N)$ with respect to the particle number. To illuminate this relationship let $\bar{\rho}$ be an electron density satisfying

$$\int d^3\mathbf{r}\ \bar{\rho}(\mathbf{r}) = N \tag{2.63}$$

and define $\Delta\rho$ as a small increment to $\bar{\rho}$ of altogether ΔN electrons. The correction Δ_{xc} is then given by [3]

$$\Delta_{\mathrm{xc}}(\mathbf{r}) = \lim_{\Delta N \to 0} [v_{\mathrm{xc}}(\mathbf{r}, N + \Delta N) - v_{\mathrm{xc}}(\mathbf{r}, N - \Delta N)] \tag{2.64}$$

and employing equation (2.42) yields

$$\Delta_{\mathrm{xc}}(\mathbf{r}) = \lim_{\Delta\rho \to 0} \left[\left. \frac{\delta E_{\mathrm{xc}}[\rho(\mathbf{r})]}{\delta\rho(\mathbf{r})} \right|_{\bar{\rho}+\Delta\rho} - \left. \frac{\delta E_{\mathrm{xc}}[\rho(\mathbf{r})]}{\delta\rho(\mathbf{r})} \right|_{\bar{\rho}-\Delta\rho} \right] . \tag{2.65}$$

When we apply the local density approximation, equation (2.62) must be extended by an additional term Δ^{LDA}, reflecting the errors introduced by the LDA compared to the exact density functional theory

$$\Delta = \Delta\epsilon^{\mathrm{LDA}} + \Delta_{\mathrm{xc}}^{\mathrm{LDA}} + \Delta^{\mathrm{LDA}} = \epsilon_{N+1}^{\mathrm{LDA}}(N) - \epsilon_{N}^{\mathrm{LDA}}(N) + \Delta_{\mathrm{xc}}^{\mathrm{LDA}} + \Delta^{\mathrm{LDA}} . \tag{2.66}$$

Since the LDA replaces the functional dependence on the density by a continuous differentiable function of the local density and thus annihilates the discontinuities of the exact exchange correlation potential $v_{\mathrm{xc}}(\mathbf{r}, N)$, the contribution $\Delta_{\mathrm{xc}}^{\mathrm{LDA}}$ vanishes and we have [5]

$$\Delta = \Delta\epsilon^{\mathrm{LDA}} + \Delta^{\mathrm{LDA}} , \tag{2.67}$$

where $\Delta\epsilon^{\mathrm{LDA}}$ is the eigenvalue gap in the local density approximation. For a multitude of semiconductors and insulators the calculated eigenvalue gap $\Delta\epsilon^{\mathrm{LDA}}$ is some ten percent smaller than the band gap Δ.

2.4 Band Structure Methods

The second Hohenberg–Kohn theorem gives rise to a variational principle for determination of the ground state of an interacting electron system. Consequently, it is reasonable to expand the single-particle wave functions from the Kohn–Sham equations (2.41) into a suitable set of basis functions and minimize the energy with respect to the coefficients of this expansion. For simplicity, the $\mathbf{k}$-point index subsequently comprises the label of the different states at a particular $\mathbf{k}$-point, i.e. the band index. Using expansion coefficients $c_i(\mathbf{k})$ and basis functions $\varphi_{\mathbf{k},i}(\mathbf{r})$ we may write

$$\chi_{\mathbf{k}}(\mathbf{r}) = \sum_i c_i(\mathbf{k}) \varphi_{\mathbf{k},i}(\mathbf{r}) . \tag{2.68}$$

As a consequence, the variational equation for the total energy functional with respect to the above expansion coefficients has the form [24]

$$\sum_{\mathbf{k},i} \left(\frac{\delta E[\rho(\mathbf{r})]}{\delta c_i^*(\mathbf{k})} + \epsilon_{\mathbf{k}} \frac{\delta \left(N - \int d^3\mathbf{r}\, \rho(\mathbf{r}) \right)}{\delta c_i^*(\mathbf{k})} \right) \delta c_i^*(\mathbf{k}) = 0 . \tag{2.69}$$

In the non-interacting case electron density and kinetic energy are given by

$$\rho(\mathbf{r}) = \sum_{\mathbf{k}} |\chi_{\mathbf{k}}(\mathbf{r})|^2 \Theta(E_F - \epsilon_{\mathbf{k}}) \tag{2.70}$$

$$T[\rho(\mathbf{r})] = \sum_{\mathbf{k}} \langle \chi_{\mathbf{k}}(\mathbf{r})| - \frac{\hbar^2}{2m} \nabla^2 |\chi_{\mathbf{k}}(\mathbf{r})\rangle \Theta(E_F - \epsilon_{\mathbf{k}}) , \tag{2.71}$$

where the Heaviside step function $\Theta(E_F - \epsilon_{\mathbf{k}})$ restricts the appearing sums to the occupied states. As the coefficients in equation (2.68) are independent of each other, the expression in the brackets of equation (2.69) vanishes for each i. Using upper relations for the electron density and kinetic energy we find the so-called secular equation [24]

$$\sum_j \left(\langle \varphi_{\mathbf{k},i}(\mathbf{r})| - \frac{\hbar^2}{2m}\nabla^2 + v_{\text{eff}}(\mathbf{r})|\varphi_{\mathbf{k},j}(\mathbf{r})\rangle - \epsilon_{\mathbf{k}}\langle\varphi_{\mathbf{k},i}(\mathbf{r})|\varphi_{\mathbf{k},j}(\mathbf{r})\rangle \right) c_j(\mathbf{k}) = 0 \,. \tag{2.72}$$

Because the solution of this linear equation system ought to be non-trivial, the determinant of the secular matrix must vanish and the single-particle energies $\epsilon_{\mathbf{k}}$ can be determined. After having established the band structure $\epsilon_{\mathbf{k}}$ the corresponding density of states can be calculated by the relation (the sum comprises all bands and **k**-points of the first Brillouin zone)

$$\text{DOS}(E) = \frac{1}{N}\sum_{\mathbf{k}} \delta(E - \epsilon_{\mathbf{k}}) \,. \tag{2.73}$$

The particular choice of basis functions in the expansion of the single-particle wave functions defines the secular matrix. Because this set of functions must be finite, for practical reasons, we cannot expect it to be complete. Consequently, the actual shape of the functions is essential for the size of the basis and the dimension of the secular matrix. If the basis functions $\varphi_{\mathbf{k},i}(\mathbf{r})$ resemble the single-particle wave functions $\chi_{\mathbf{k}}(\mathbf{r})$ well, a large fraction of the single-particle Hilbert space is spanned. Hence a smaller basis set is sufficient for a suitable expansion and the computational effort for solving the secular equation is reduced.

Because of the periodicity of a crystal one may suggest that an appropriate set of basis functions is given by the plane waves

$$\varphi_{\mathbf{k},\nu}(\mathbf{r}) = \exp(i(\mathbf{k} + \mathbf{K}_\nu)\mathbf{r}) \,, \tag{2.74}$$

which offer the additional advantage to be orthogonal. But the shapes of plane waves do not reflect the characteristics of the crystal potential, which results from a superposition of atomic potentials and hence is similar to the bare nuclear Coulomb potential close to the atomic sites, whereas it is rather flat in the interatomic regions. As a consequence, the single-particle wave functions will show oscillations near the nuclei and become smoother in interstitial areas. A modeling of such functions by means of plane waves would need a large basis set – making the choice of a plane wave basis unfavorable. A better adjustment of the basis functions to the characteristics of the crystal potential is achieved within the orthogonalized plane wave method. In this scheme one starts out from a plane wave basis but orthogonalizes each wave function to the strongly bound core states $|\phi_c\rangle$

$$\varphi_{\mathbf{k},\nu}(\mathbf{r}) = |\mathbf{k} + \mathbf{K}_\nu\rangle - \sum |\phi_c\rangle\langle\phi_c|\mathbf{k} + \mathbf{K}_\nu\rangle \,. \tag{2.75}$$

Thus each orthogonalized plane wave consists of a long range part due to the underlying plane wave as well as of strongly oscillating contributions near the nuclei. By reformulating the orthogonalization of the basis functions to the core states in terms of an additional potential seen by the valence electrons we arrive at the so-called pseudopotential methods. Another method to match the basis functions entering the variational principle to the details of the crystal potential arises from the muffin-tin approximation given by J. C. Slater

[25]. Here one assumes the potential to be spherically symmetric within particular non-overlapping spheres and constant in the remaining regions. The effective single particle potential v_σ^{MT} entering the Schrödinger equation

$$\left(-\frac{\hbar^2}{2m}\nabla^2 + v_\sigma^{\mathrm{MT}}(\mathbf{r}) - \epsilon\right)\chi_\sigma(\epsilon,\mathbf{r}) = 0 \tag{2.76}$$

hence takes the form

$$v_\sigma^{\mathrm{MT}}(\mathbf{r}) = v^{\mathrm{MTZ}}\Theta_{\mathrm{int}} + \sum_{\mu,\nu} v_{\nu,\sigma}^{\mathrm{MT}}(\mathbf{r}_{\mu,\nu})\Theta_{\mu,\nu}\,. \tag{2.77}$$

In this equation v^{MTZ} represents the constant interstitial potential, the muffin-tin zero, and the step functions Θ_{int} and $\Theta_{\mu,\nu}$ delimit the contributions to the interstitial regions and to the atomic spheres centered at sites $\mathbf{R}_{\mu,\nu}$, respectively. Moreover, the definitions $\mathbf{r}_{\mu,\nu} = \mathbf{r} - \mathbf{R}_{\mu,\nu}$ and $\mathbf{R}_{\mu,\nu} = \mathbf{R}_\mu + \mathbf{t}_\nu$ with the lattice vector $\mathbf{R}_\mu$ and the position $\mathbf{t}_\nu$ of the nucleus ν in the unit cell have been applied. Plane waves might be used to describe the wave function in the interstitial regions. Within the atomic spheres the radial Schrödinger equation can be solved numerically and by matching the radial functions continuously to the plane waves at the sphere boundaries we can construct augmented plane waves. Considering not only the spherically symmetric part but also the unrestricted potential in the secular matrix we establish a full potential method. The latter allows for a very accurate calculation – even of elastic properties. Hence the muffin-tin potential approximates the crystal potential well and yields adequate basis functions.
As we have discussed, it is generally disadvantageous to deal with a plane wave basis. Despite some mathematical simplicity and the motivation by the crystalline periodicity it is preferable to put more emphasis on the single atom and the particular shape of its nuclear potential. Therefore one may opt for spherical waves as solutions of the Schrödinger equation in the regions with the constant potential. Spherical waves are products of spherical harmonics and spherical Hankel, Neumann, and Bessel functions. Augmenting the spherical waves growing out of the interatomic regions to the solutions of the radial Schrödinger equation in the muffin-tin spheres, one obtains the Korringa-Kohn-Rostoker method. In the augmented plane wave approach and in the Korringa-Kohn-Rostoker method the solutions of the Schrödinger equation within the muffin-tin spheres are energy dependent functions. Therefore the basis functions are likewise energy dependent, which strongly increases the computational effort for solving the secular equation. The energy dependence of the basis functions is actually weak in the energy region relevant for valence electrons. A Taylor series expansion (up to the first order in the energy dependence) leads to linear counterparts of both the augmented plane wave and the Korringa-Kohn-Rostoker method: the linear augmented plane wave and the linear muffin-tin orbital method [26].
In contrast to the muffin-tin approximation and its restriction to non-overlapping spheres, the related atomic sphere approximation (ASA) requires the volumes Ω_i of the atomic spheres to sum up to the volume of the unit cell Ω_{uc}

$$\sum_i \Omega_i = \Omega_{\mathrm{uc}}\,. \tag{2.78}$$

The approximation traces back to O. K. Andersen [26] and offers the advantage of formally eliminating the interstitial regions of the muffin-tin scheme by blowing up the muffin-tin spheres. An approach usually applying the ASA is the augmented spherical wave (ASW) method as developed by A. R. Williams, J. Kübler, and C. D. Gelatt Jr. [27]. The scheme is closely related to the linear muffin-tin orbital method since they both adopt basis functions tracing back to spherical waves. A detailed description of the fundamental notions of the ASW method was given by V. Eyert [24]. The advantage of this technique is the combination of a linear method with the benefits of a minimal basis set applying spherical waves, which allows investigation of compounds with complex crystal structures and intuitive interpretion of the calculated electronic properties in terms of the atomic orbitals. As the linear muffin-tin orbital approach, the ASW method suppresses the energy dependence of the wave functions and applies radial functions for a fixed energy value, which is usually $-0.015\,\mathrm{Ryd}$ [18, 28].

Beyond the calculation of band structures $\epsilon_{\mathbf{k}}$ and densities of states $\mathrm{DOS}(E)$ the investigations of the following chapters will require the detailed analysis of chemical bonding. In this context, the concept of the crystal orbital overlap population (COOP) is a tool for studying the electronic and chemical properties of materials, which was introduced by R. Hoffmann [29] especially to allow for the analysis of chemical bonding. In short, the COOP is based on expectation values of operators consisting of the non-diagonal elements of the overlap population matrix $c_i^*(\mathbf{k})O_{ij}c_j(\mathbf{k})$. Here

$$O_{ij} = \langle \varphi_{\mathbf{k},i}(\mathbf{r}) | \varphi_{\mathbf{k},j}(\mathbf{r}) \rangle \tag{2.79}$$

is an element of the overlap matrix of the basis functions and the quantities $c_i(\mathbf{k})$ are the expansion coefficients from equation (2.68). The partial COOP is given by [30]

$$\mathrm{COOP}_{ij}(E) = \sum_{\mathbf{k}} \mathrm{Re}\left(c_i^*(\mathbf{k})c_j(\mathbf{k})\right) O_{ij}\delta(E-\epsilon_{\mathbf{k}})\,. \tag{2.80}$$

Here the sum comprises all bands and $\mathbf{k}$-points in the first Brillouin zone. Remember that the band index was included in the index $\mathbf{k}$. Positive contributions to the COOP indicate bonding states, whereas negative terms in equation (2.80) point to antibonding states. While the COOP is able to describe the pure bonding character, it cannot analyze quantitatively the contribution of the single bonds to the total energy. Therefore one defines the crystal orbital Hamilton population (COHP) [31]

$$\mathrm{COHP}_{ij}(E) = \sum_{\mathbf{k}} \mathrm{Re}\left(c_i^*(\mathbf{k})c_j(\mathbf{k})\right) H_{ij}\delta(E-\epsilon_{\mathbf{k}}) \tag{2.81}$$

with the Hamilton matrix elements

$$H_{ij} = \langle \varphi_{\mathbf{k},i}(\mathbf{r}) | H | \varphi_{\mathbf{k},j}(\mathbf{r}) \rangle\,. \tag{2.82}$$

Equations (2.80) and (2.81) allow for introducing the covalence energy E_{COV}

$$E_{\mathrm{COV},ij}(E) = \sum_{\mathbf{k}} \mathrm{Re}\left(c_i^*(\mathbf{k})c_j(\mathbf{k})\right) \left[H_{ij} - O_{ij}\frac{1}{2}(H_{ii}+H_{jj})\right] \delta(E-\epsilon_{\mathbf{k}})\,. \tag{2.83}$$

$E_{\mathrm{COV,ij}}(E)$ may be interpreted as a kind of (quantitative) covalent bond strength of the pairs of orbitals i and j. Because COOP and COHP change their sign at the same energy when moving from bonding to antibonding states and vice versa, we can equally well use the sign of the covalence energy E_{COV} to describe the bonding character. Bonding of the respective orbitals is indicated by negative energies, whereas antibonding corresponds to positive values.
The total covalence energy is evaluated as the sum over all non-diagonal elements

$$E_{\mathrm{COV}}(E) = \sum_{i \neq j} E_{\mathrm{COV},ij}(E) \,. \tag{2.84}$$

In order to obtain an impression of the chemical stability of a crystal we introduce the integrated total covalence energy

$$\tilde{E}_{\mathrm{COV}}(E) = \int_{-\infty}^{E} dE' \, E_{\mathrm{COV}}(E') \,. \tag{2.85}$$

Accordingly, the integrated partial covalence energies $\tilde{E}_{\mathrm{COV},ij}(E)$ provide practical measures for the chemical stability of certain bonds.

Chapter 3

Vanadium Dioxide and Sesquioxide: Structural and Electronic Transitions

To prepare for the investigation of the vanadium Magnéli phases a very detailed analysis of vanadium dioxide and sesquioxide is required. As functions of temperature and doping both compounds undergo metal-insulator transitions (MITs) combined with transformations of the crystal structure. The mechanisms of these transitions have been investigated intensively for decades but are still a matter of dispute. Central findings are reviewed in the articles of J. B. Goodenough [32], W. Brückner *et al.* [33], and M. Imada *et al.* [34].

3.1 Crystal Structure of VO_2

In the following, we analyze the crystal structure of VO_2. This is useful not only in order to understand the results of the electronic structure calculations discussed subsequently but also allows for a comprehensive insight into the crystal structures of the whole Magnéli series. Above 340 K VO_2 crystallizes in the rutile structure, which traces back to a simple tetragonal lattice with space group $P4_2/mnm$ (D_{4h}^{14}), see figure 3.1. Using single crystal experiments D. B. McWhan *et al.* [35] determined the tetragonal lattice constants $a_R = 4.5546$ Å and $c_R = 2.8514$ Å and the positional parameters. They reported the vanadium atoms to occupy the Wyckoff positions (2a): (0,0,0) and (1/2,1/2,1/2). In addition, the oxygen atoms are located at the Wyckoff positions (4f) with the parameter $u = 0.3001$: $(\pm u, \pm u, 0)$ and $(1/2 \pm u, 1/2 \pm u, 1/2)$. The primitive translations of the simple tetragonal (rutile) lattice are defined as

$$\mathbf{a}_R = \begin{pmatrix} a_R \\ 0 \\ 0 \end{pmatrix}, \quad \mathbf{b}_R = \begin{pmatrix} 0 \\ a_R \\ 0 \end{pmatrix}, \quad \mathbf{c}_R = \begin{pmatrix} 0 \\ 0 \\ c_R \end{pmatrix}. \tag{3.1}$$

With respect to the metal atoms the crystal structure of VO_2 can be described in terms of a regular body-centered tetragonal lattice. Each of the vanadium sites is surrounded by an oxygen octahedron. As is demonstrated in figure 3.1, the respective octahedra belonging to the corners and to the center of a rutile unit cell are rotated by 90° with respect to the

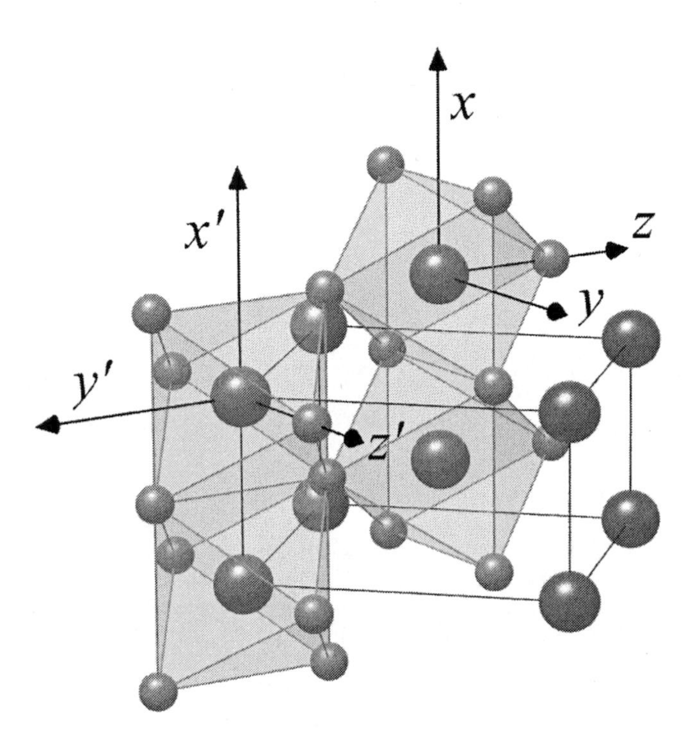

Figure 3.1: *Crystal structure of metallic* VO_2. *The inserted cuboid determines the unit cell of the rutile structure with the c-axis oriented upwards. Large and small spheres represent vanadium and oxygen atoms, respectively. The vanadium atoms form infinite chains along the rutile c-axis. Furthermore, each vanadium site is surrounded by an oxygen octahedron. The latter are oriented uniformly in each vanadium chain but rotated by 90° with respect to the rutile c-axis among neighbouring chains. Hence they establish local coordinate systems.*

rutile c-axis. This reduces the lattice symmetry from body-centered tetragonal to simple tetragonal and causes the unit cell to contain two formular units. Equally adjusted oxygen octahedra form chains along the rutile c-axis. Within the chains neighbouring octahedra share edges, whereas corner and center chains themselves are connected via corners. Next to the vanadium filled oxygen octahedra the crystal structure of VO_2 comprises the same amount of empty octahedra, which set up chains along the rutile c-axis as well. Hence the crystal structure may be described as a regular three dimensional network of oxygen octahedra partially filled by vanadium atoms. Each octahedron has orthorhombic symmetry, but the deviations from tetragonal and even cubic symmetry are small, which allows for discussion in terms of the latter. The distortion of the oxygen octahedron yields two different V-O distances [36]. More precisely, we find the apical distance between metal and

oxygen atoms with identical z_R values in the rutile lattice. Hence it appears twice within each octahedron. In contrast, the equatorial distance is found four times connecting the vanadium atom to the remaining oxygen sites shifted by $\pm 1/2 \cdot \mathbf{c}_R$.
For the discussion of the electronic structure of VO_2 it is necassary to define local coordinate systems centered at the metal sites. Due to the different orientation of the oxygen octahedra at the corners and the center of the rutile unit cell two different reference systems are required, see figure 3.1. In both cases the local z-axis is oriented along the apical axis of the local ocahedron, which is either the (110) or the (1$\bar{1}$0) direction. In comparison to the traditional alignment of the local x-axis as well as the local y-axis parallel to the metal-ligand bonds, these axes are rotated by 45° with respect to the local z-axis. Hence they are parallel and perpendicular to the rutile c-axis, respectively. This definition of a local rotated reference frame below will be used not only for rutile vanadium dioxide but also for compounds with related crystal structures. For an analogous representation in all these cases, the pseudorutile axes a_{prut}, b_{prut} and c_{prut} are introduced. They are identical to the original rutile axes in the rutile case and still refer to them otherwise.
The previous definition of a local coordinate system will allow for a detailed analysis of the correlations between structural features and electronic properties. For this purpose figure 3.2 depicts the angular parts of the metal d orbitals relative to the local reference frame of the central metal atom. The cubic part of the crystal field splitting results in a separation of the d orbitals in threefold degenerate t_{2g} and twofold degenerate e_g states. The former comprise the $3d_{x^2-y^2}$, $3d_{xz}$, and $3d_{yz}$ states, whereas the latter consist of the $3d_{3z^2-r^2}$ and $3d_{xy}$ orbitals. Both e_g orbitals point directly towards the oxygen atoms of the local octahedron. The $d_{x^2-y^2}$ orbital is oriented along the rutile c-axis (i.e. the local x-axis) and the local y-axis. It is contained in the basal plane of the oxygen octahedron, pointing at its edges. The d_{xz} and d_{yz} orbitals instead are directed towards the faces of the octahedron. From figure 3.2 it becomes obvious that $d_{x^2-y^2}$ and d_{xz} states mediate σ and π-type V-V overlap along the rutile c-axis, respectively. Due to the aforementioned 45° rotation of the local coordinate system the $d_{x^2-y^2}$ and d_{xy} orbitals are interchanged compared to the standard notation. Furthermore, the lobes of the d_{yz} orbital point perpendicular to the rutile c-axis. Due to the measured length ratio $c_R/a_R = 0.6260$ of the crystal axes the latter orbital exhibits a significantly reduced σ-type overlap compared to the $d_{x^2-y^2}$ states. While both the $d_{x^2-y^2}$ and d_{yz} orbitals connect vanadium sites separated by tetragonal lattice vectors, coupling between metal sites located at the center and the corners of a unit cell is mediated by d_{xz} orbitals, which exhibit a remarkable overlap with the $d_{x^2-y^2}$ states belonging to the vanadium atoms of neighbouring octahedral chains.
Many vanadium oxides show a temperature induced MIT as observed in V_2O_3 and the Magnéli phases V_nO_{2n-1} ($3 \leq n \leq 9$), for instance. The transition of VO_2 takes place at about 340 K, accompanying a structural transition from the high temperature rutile (R phase) to a low temperature monoclinic structure (M_1 phase). Two additional insulating phases occur on the application of uniaxial stress along the rutile (110) axis or on doping with some percent of chromium or aluminium. The generalized phase diagram is shown in figure 3.3 [37–39]. For stoichiometric vanadium dioxide at ambient pressure the MIT is of first order [40]. Exactly the same applies to the MIT of doped VO_2 from the rutile to the monoclinic M_2 phase. In the latter case further lowering of the temperature yields a transition into the triclinic T phase and another transition into the M_1 phase for suffi-

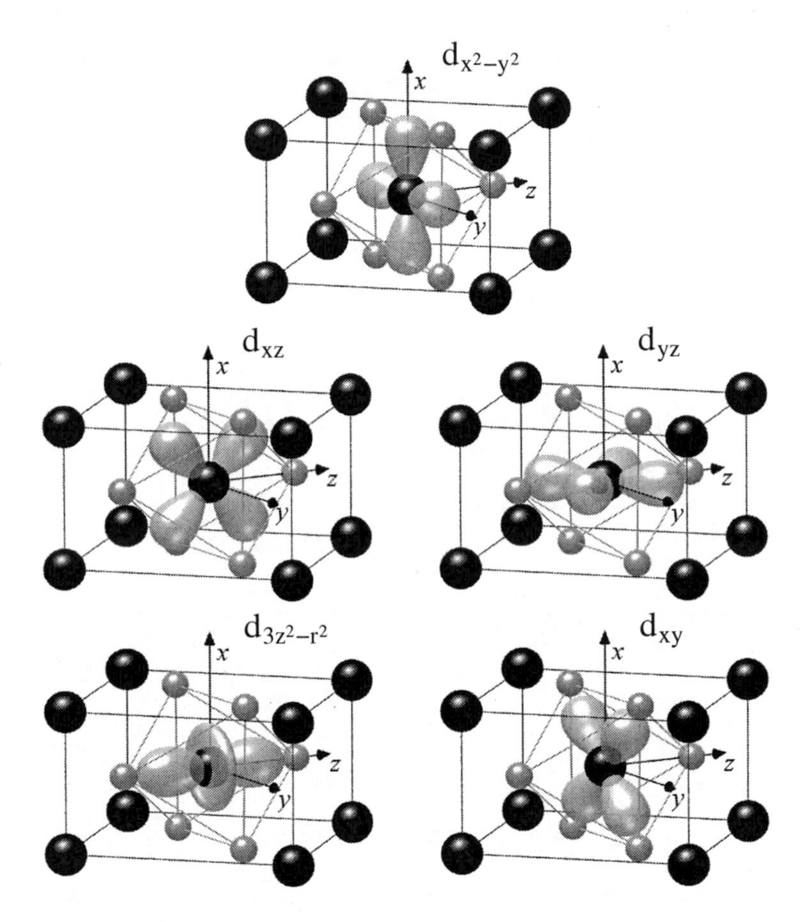

Figure 3.2: *Angular parts of the transition metal 3d states with respect to the local coordinate system of the central metal atom in the displayed rutile unit cell. Note that the rutile c-axis is oriented upwards. The t_{2g} manifold, resulting from the crystal field splitting of the metal states, comprises the $d_{x^2-y^2}$, d_{xz} and d_{yz} orbital, whereas the e_g manifold comprises both $d_{3z^2-r^2}$ and d_{xy} states. While the $d_{x^2-y^2}$ lobes point at edges of the local octahedron, the d_{xz} and d_{yz} orbitals are directed towards faces. Both e_g orbitals point at oxygen atoms.*

ciently small chromium concentrations. For the insulating phases figure 3.3 additionally shows distortion patterns of the metallic vanadium chains along the rutile *c*-direction. In the M_1 phase both metal-metal pairing and zigzag-type lateral displacements affect each chain. In contrast, in the M_2 phase half of the chains dimerize and the other half show zigzag-type deviations. Finally, the T phase is intermediate since the dimerized M_2 chains

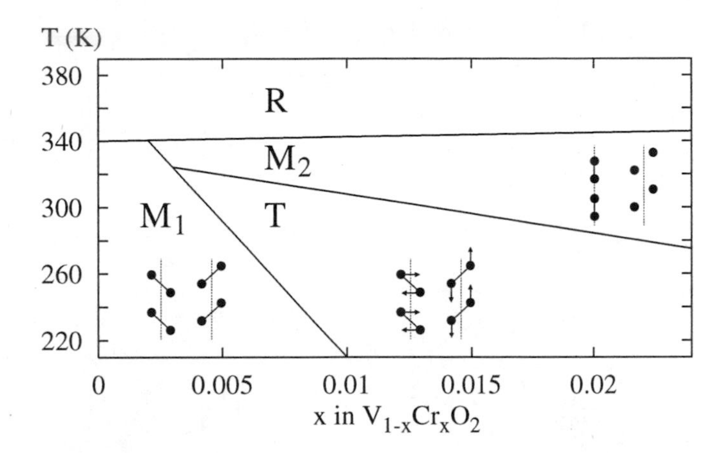

Figure 3.3: *Generalized phase diagram of $V_{1-x}Cr_xO_2$ [37, 39]. Besides the metallic rutile (R) phase three insulating modifications of $V_{1-x}Cr_xO_2$ are observed: at ambient pressure pure VO_2 enters a monoclinic (M_1) phase below 340 K. Under uniaxial stress or on doping a second monoclinic (M_2) and a triclinic (T) phase are observed. The associated distortion patterns of the metallic vanadium chains (viewed along a_{prut}) are depicted as insets in the insulating phases. Bars connecting vanadium sites clarify metal-metal pairing. The arrows in the T phase indicate the displacement of the vanadium atoms on increasing temperature.*

gradually start to tilt and the zigzag chains evolve a dimerization until eventually the M_1 structure is reached.

In order to understand the structural changes at the MIT of (pure) VO_2 in more detail we consider the monoclinic M_1 structure based on a simple monoclinic lattice with space group $P2_1/c$ (C_{2h}^5), see figure 3.4. J. M. Longo and P. Kierkegaard [41] reported the values $a_{M_1} = 5.7517$ Å, $b_{M_1} = 4.5378$ Å, $c_{M_1} = 5.3825$ Å, and $\beta_{M_1} = 122.646°$ for the lattice constants and the monoclinic angle, respectively. A unit cell comprises four formula units. The vanadium atoms and both crystallographically inequivalent oxygen sites are located at the Wyckoff positions (4e): $\pm(x, y, z)$ and $\pm(x, 1/2 - y, 1/2 + z)$. Here the coordinates refer to the standard choice

$$\mathbf{a}_{M_1} = \begin{pmatrix} 0 \\ 0 \\ -a_{M_1} \end{pmatrix}, \quad \mathbf{b}_{M_1} = \begin{pmatrix} -b_{M_1} \\ 0 \\ 0 \end{pmatrix}, \quad \mathbf{c}_{M_1} = \begin{pmatrix} 0 \\ c_{M_1} \sin\beta_{M_1} \\ -c_{M_1} \cos\beta_{M_1} \end{pmatrix} \tag{3.2}$$

for the primitive translations of the simple monoclinic lattice. The atomic positions in the monoclinic cell are determined by the parameters $(x, y, z)_V = (0.23947, 0.97894, 0.02646)$, $(x, y, z)_{O1} = (0.10616, 0.21185, 0.20859)$, and $(x, y, z)_{O2} = (0.40051, 0.70258, 0.29884)$. A comparison of figures 3.1 and 3.4 reveals close relations between the monoclinic M_1 and the high temperature rutile structure. Two rutile unit cells added to figure 3.4 illustrate the description of the monoclinic cell in terms of a rutile cell doubled along c_{prut}.

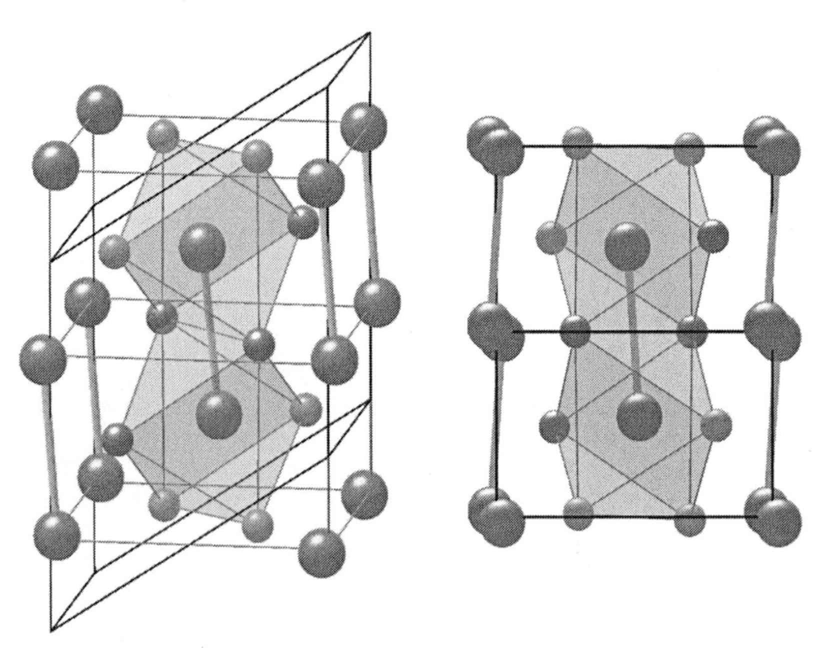

Figure 3.4: *Crystal structure of insulating VO_2. The three dimensional view displays one monoclinic unit cell. Two neighbouring rutile cells are included to illustrate the connection to the rutile structure of the metallic phase. The orientation of the crystal is the same as in figure 3.1. Large and small spheres denote vanadium and oxygen atoms, respectively. Bars connecting vanadium sites stress the observed metal-metal pairing along the vertical rutile c-axis. Furthermore, zigzag-type displacements of the vanadium atoms along the diagonal of the rutile basal plane are visible. A side view along a_{prut} demonstrates the displacements in more detail. Due to the side projection the oxygen octahedra form hexagonal structures.*

Despite this similarity we find distinct differences when comparing the structures of the high and the low temperature phase. First, a characteristic metal-metal pairing along the c_{prut}-axis is present in the monoclinic crystal structure, which modifies the V-V distances (2.851 Å in the metallic configuration) and gives rise to alternating values of 2.619 Å and 3.164 Å. Second, in the low temperature phase zigzag-type in-plane displacements of the metal atoms parallel to the local z-axes, i.e. along the diagonal of the rutile basal plane, evolve. The shift direction alternates along both a_{prut} and c_{prut} but not along b_{prut}. Due to the zigzag-type distortions two different apical V-O bond lengths of 1.77 Å and 2.01 Å are observed. In addition, the metal-metal pairing results in two short (1.86 Å and 1.89 Å) and two long (2.03 Å and 2.06 Å) equatorial distances in each oxygen octahedron. Third, a lattice strain is present in the monoclinic case; thus the ratios $c_{M_1} \sin \beta_{M_1} / b_{M_1} = 0.9988$ as well as $-2c_{M_1} \cos \beta_{M_1} / a_{M_1} = 1.0096$ differ from unity. In terms of the rutile lattice the primitive translations of the monoclinic crystal are approximated as

$$\mathbf{a}_{M_1} \approx \begin{pmatrix} 0 \\ 0 \\ -2c_R \end{pmatrix}, \quad \mathbf{b}_{M_1} \approx \begin{pmatrix} -a_R \\ 0 \\ 0 \end{pmatrix}, \quad \mathbf{c}_{M_1} \approx \begin{pmatrix} 0 \\ a_R \\ c_R \end{pmatrix}. \tag{3.3}$$

The demonstrated instability of the rutile crystal structure is found not only in the case of VO_2 but also applies to many other early transition metal oxides as for instance MoO_2 and NbO_2. For these compounds the relations of the structural distortions to the changes in the electronic structure have been investigated extensively [42, 43]. The materials are characterized by metal-metal pairing along certain chains as well as by lateral zigzag-type displacements in the low temperature phase. Despite all the similarities a striking difference regarding the in-plane distortions of the metal chains and the modifications of the surrounding oxygen octahedra distinguishes monoclinic VO_2 from MoO_2. While for the latter material the octahedra mainly follow the in-plane molybdenum shifts, they almost stay at their original positions in the case of VO_2. Hence the lateral displacements cause remarkable shifts of the metal atoms relative to the octahedra only for vanadium dioxide. Naturally, a relative movement must affect the bonding between the V $3d$ and O $2p$ states since it influences the mutual overlap of the orbitals. Therefore an antiferroelectric distortion of the whole VO_6 octahedra is inherent in the monoclinic (M_1) phase of VO_2. From the different behaviour of the oxygen octahedra in VO_2 and MoO_2 it was concluded that the antiferroelectric distortion does not explain the tendency of the transition metal oxides to form warped variants of the ideal rutile structure. For the same reason this mechanism must be excluded as a possible driving force of MITs connected to the destabilization of the rutile crystal structure [44].

3.2 Electronic Properties of VO_2

The electronic structure of VO_2 has been attracting interest mainly due to the MIT the stoichiometric compound undergoes at a temperature of roughly 340 K. Experimentally, in transport measurements one observes a sharp drop in resistivity extending over several orders of magnitude [40]. The transition occurs simultaneously to a transformation from the high temperature rutile to the low temperature monoclinic structure. From the theoretical point of view many models, ranging from Peierls [45, 46] to Mott–Hubbard [47, 48] schemes, were proposed to understand its origin. To different degrees the approaches consider lattice instabilities, electron-phonon interaction, as well as electronic correlations as possible driving forces.
The important role of the lattice degrees of freedom for the stabilization of the different phases of VO_2 together with the symmetry change in the crystal structure was taken as evidence for strong electron-phonon coupling and a lattice softening close to the transition [35]. This argumentation is encouraged by the comparison of the elastic properties of rutile and monoclinic vanadium dioxide via ultrasonic microscopy [49]. The latter investigation reveals a strong elastic anisotropy in the metallic phase, which disappears almost completely when the insulating phase is entered. The electronic structure of metallic VO_2 has been probed by optical measurements revealing the lowest unfilled V $3d$ levels 2.5 eV

above the top edge of the oxygen $2p$ bands [50]. Ultraviolet as well as x-ray photoelectron spectroscopy show an approximately 8.5 eV wide occupied band directly below the Fermi energy [51]. Going into more detail this valence band splits up into low and high binding regions with widths of about 1.5 eV and 6 eV, respectively. While the low binding contributions are attributed to the V $3d$ states, the broader part of the valence band mainly results from the O $2p$ states. According to oxygen K-edge x-ray absorption spectroscopy additional unoccupied V $3d$ bands extend from the Fermi energy to 1.7 eV and from 2.2 eV to 5.2 eV [52]. For the monoclinic modification of VO_2 photoelectron spectroscopy displays a sharpening accompanied by an energetical downshift of the occupied V $3d$ bands. S. Shin *et al.* [53] found a band gap of about 0.7 eV in the insulating phase.
A very frequently discussed energy band scheme for both the metallic and the insulating phase was proposed by J. B. Goodenough [32]. Starting with electrostatic considerations he placed the O $2p$ levels in a first step well below the V $3d$ states. Due to the nearly octahedral crystal field the latter split up into the lower t_{2g} and the higher e_g levels. In this context the t_{2g} bands are located in the vicinity of the Fermi energy and additionally split up into the $d_{\|}$ state, oriented along the rutile c-axis, and the remaining π^* states. When entering the monoclinic phase the metal-metal pairing in the vanadium chains parallel to the rutile c-axis causes the $d_{\|}$ band to split up into filled bonding and empty antibonding states. Furthermore, as a consequence of the antiferroelectric zigzag-type displacements of the vanadium atoms, the π^* bands reveal a shift to higher energies. S. Shin *et al.* [53] report the $d_{\|}$ band splitting to amount to about 2.5 eV and the π^* bands to rise by 0.5 eV.
A. Zylbersztejn and N. F. Mott [54] proposed an MIT mechanism based on the presence of strong electron-electron correlations. According to the above authors especially the one-dimensional $d_{\|}$ band is affected by correlations rather than by electron-lattice interaction. However, in the metallic case the correlations are efficiently screened by the π^* electrons. The screening diminishes below the phase transition because the π^* bands are subject to energetical upshift because of antiferroelectric-like displacements of the vanadium atoms. Therefore the narrow $d_{\|}$ bands become susceptible to strong Coulomb repulsions and thus undergo a Mott-transition, which opens up a correlation gap.
'First principles' molecular dynamics calculations by R. M. Wentzcovitch *et al.* [55] point to the predominant influence of the lattice degrees of freedom. The authors use a variable cell shape approach allowing for the simultaneous relaxation of the atomic positions and primitive translations of the lattice. Starting with different intermediate structures they find the monoclinic M_1 configuration to be the most stable. Furthermore, they obtain a slightly metastable rutile solution with a total energy per formula unit 54 meV higher than the monoclinic energy. In each case the calculated structural parameters agree well with experimental findings. R. M. Wentzcovitch *et al.* miss the insulating gap for the monoclinic configuration but report a semimetallic behaviour with a band overlap of 0.04 eV. This result might be attributed to the known failure of the local density approximation in reproducing optical band gaps. It was presumed that a strengthening of the V-V bonds opens the gap and VO_2 thus can be regarded a band insulator. In an LDA+U approach W.-D. Yang [56] reproduced the correct optical band gap. It is a matter of dispute if the MIT of VO_2 can be reproduced by the LDA+DMFT approach [57,58].
Further experimental and theoretical results for vanadium dioxide were summarized by V. Eyert [44,59]. Concerning the mechanism of the MIT in VO_2 the electronic structure

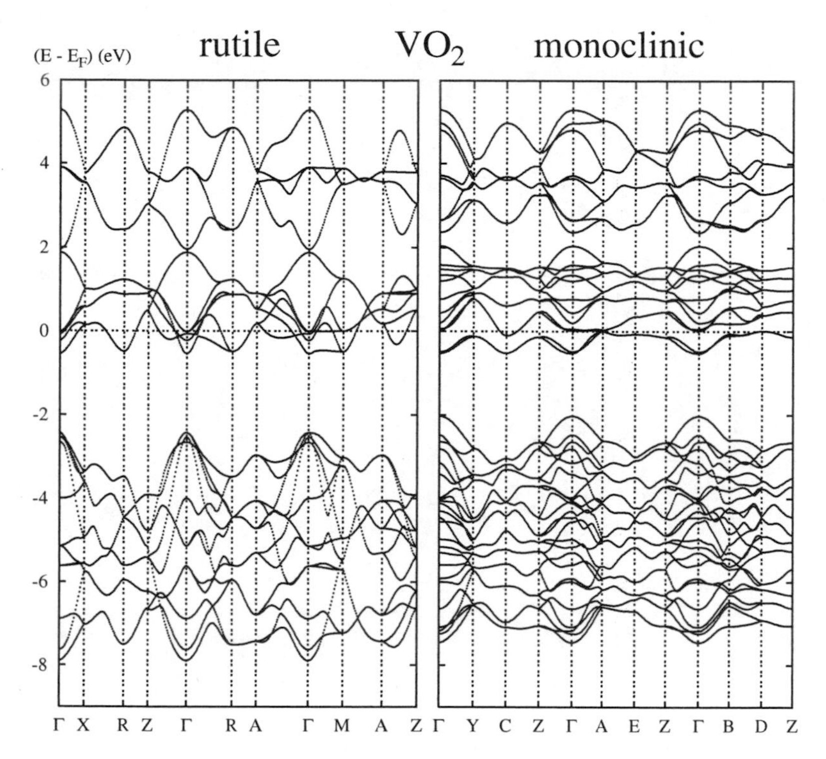

Figure 3.5: *Electronic bands of rutile and monoclinic VO_2 shown along selected symmetry lines in the first Brillouin zone of the simple tetragonal and the simple monoclinic lattice, respectively. Regarding the definition of the different symmetry points compare figure 3.6.*

calculations presented in the above articles give strong hints at a Peierls instability of the one-dimensional $d_{\parallel}$ band embedded in the background of the remaining V $3d$ t_{2g} states. Because the results for VO_2 will form the basis of investigating the class of the vanadium Magnéli phases in the next chapter, we review them in some detail. Figure 3.5 gives the electronic states growing out of an LDA calculation for rutile and monoclinic (M_1) VO_2. The bands are shown along selected high symmetry lines in the first Brillouin zone of the simple tetragonal and the simple monoclinic lattice. In order to establish the nomenclature of the high symmetry points the Brillouin zones are displayed in figure 3.6, compare C. J. Bradley and A. P. Cracknell [60].

In figure 3.5 one easily identifies three groups of bands for both the rutile and the monoclinic phase of VO_2. In the rutile case they are located in the energy ranges from $-7.9\,\mathrm{eV}$ to $-2.4\,\mathrm{eV}$, from $-0.6\,\mathrm{eV}$ to $1.9\,\mathrm{eV}$, and from $2.0\,\mathrm{eV}$ to $5.3\,\mathrm{eV}$. This arrangement reflects the expectations due to the molecular orbital picture of J. B. Goodenough. The hybridization between the O $2p$ and the V $3d$ orbitals leads to σ- and π-type overlap yielding bonding

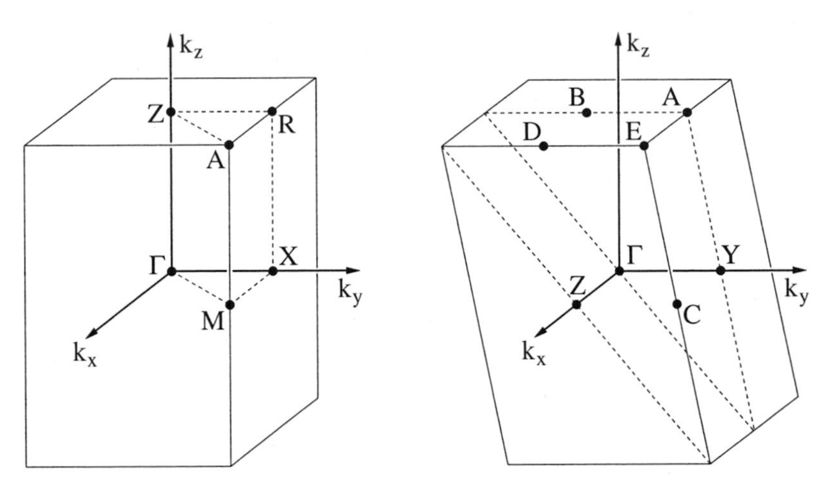

Figure 3.6: *First Brillouin zones of the simple tetragonal (left) and the simple monoclinic (right) lattice. Dashed lines mark the irreducible wedge and labels denote symmetry points.*

σ and π as well as antibonding σ^* and π^* states. Because the overlap is increased for the σ-type states their bonding-antibonding splitting accordingly is larger, which establishes the energetical order σ-π-π^*-σ^* for the molecular orbitals. While the bonding states are filled and predominately reveal O $2p$ character, the antibonding bands are found in the vicinity of the Fermi level; they are dominated by the V $3d$ states. Because of the nearly ideal cubic octahedral surrounding of the vanadium atoms, the σ^* and π^* states show e_g^σ and t_{2g} symmetry, respectively. From the three t_{2g} orbitals particularly the $d_{x^2-y^2}$ states are influenced by strong V-V interactions parallel to the rutile c-axis, compare figure 3.2. Consequently, their symmetry is b_{1g} and due to overlap parallel to the vanadium chains they are called $d_\parallel$ states. In the following, we will denote the remaining t_{2g} orbitals as e_g^π states. The frequently applied notation π^* causes confusion with the identical labeling of the t_{2g} states as a whole. In figure 3.5 the first group of bands comprises 12, the second 6, and the third 4 individuals. Bands are counted most easily along lines where they are twofold degenerate, for example along X-R. Notedly, there are two formula units per unit cell. Adopting the molecular orbital point of view one interprets bands between $-7.9\,\text{eV}$ and $-2.4\,\text{eV}$ as σ and π states. The π^* and σ^* states occupy the regions from $-0.6\,\text{eV}$ to $1.9\,\text{eV}$ and from $2.0\,\text{eV}$ to $5.3\,\text{eV}$, respectively.
The first row of figure 3.7 shows partial V $3d$ and O $2p$ densities of states (DOS) calculated for the rutile and the monoclinic crystal structure in the same energy range used in figure 3.5. As a matter of fact, the three groups of bands appearing in figure 3.5 are identified in the DOS. Confirming the prediction of the molecular orbital picture the lowest group primarily traces back to the O $2p$ states, whereas the other groups mainly originate from the V $3d$ orbitals. Nevertheless, we can observe non-negligible contributions of vanadium and oxygen in the energy regions dominated by the respective other states. They are due

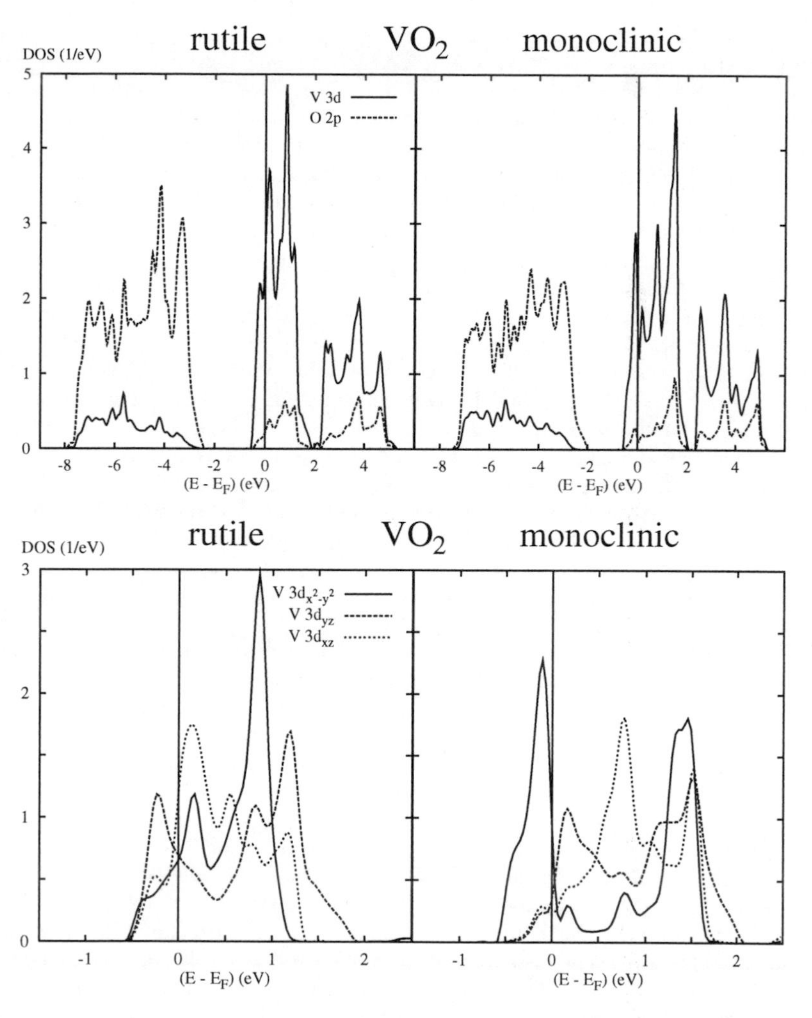

Figure 3.7: *Partial densities of states (DOS) per vanadium atom of rutile and monoclinic VO_2. In the first row the partial V 3d and O 2p DOS resulting from the crystal structures of metallic (left) and insulating (right) VO_2 are compared. In the second row the partial V 3d t_{2g} DOS is depicted in detail. Here the orbitals refer to the local rotated reference frame.*

to the hybridization between V $3d$ and O $2p$ states, which naturally is larger in the case of a σ-type overlap. Hence the mutual contributions are increased in the σ and σ^*-areas compared to the respective π and π^*-ranges. The second row of figure 3.7 explicitly shows the partial V $3d$ t_{2g} DOS for both crystal structures. Thereby the presentation of the data refers to the local rotated reference frame as discussed in the preceding section. It is worth mentioning that in the chosen energy interval hardly any contributions due to the V $3d$ e_g^σ orbitals can be observed. Hence the separation in t_{2g} and e_g^σ states is almost perfect. Obviously, the mixing of these configurations is a good measure for the distortions the oxygen octahedra exhibit. The distinct difference between the t_{2g} symmetry components in figure 3.7 reflects the orthorhombic coordination of the vanadium sites. As a consequence of the metal-metal bonding perpendicular to the rutile c-axis the partial d_{yz} DOS reveals a pronounced double peak structure. The separation of the two peaks in the rutile $d_{x^2-y^2}$ DOS is significantly smaller but still indicative of V-V overlap along the vanadium chains. In the next step we concentrate on the modifications of the band structure and the partial DOS when changing the crystal structure from rutile to monoclinic, see the right sides of figures 3.5 and 3.7. The same groups of bands can be identified as for rutile VO_2. However, due to the fact that the monoclinic unit cell comprises four formula units now 24 oxygen dominated bands are observed well below the Fermi energy. Furthermore, two groups of 12 and 8 bands, which mainly trace back to the V $3d$ states, are located at and above the Fermi energy, respectively. Admixtures of V $3d$ and O $2p$ in the regions where the respective other bonding partner dominates again are due to the hybridization. In accordance with the molecular orbital picture the energetical separation between the V $3d$ and the O $2p$ derived bands is somewhat smaller than for rutile VO_2. Despite the impressive similarity of the findings a more detailed analysis shows distinct differences. First of all, two pronounced peaks evolve in the partial V $3d$ t_{2g} DOS at energies of about -0.2 eV and 1.4 eV. Because spectral weight is shifted from the center of the t_{2g} group of states to its edges, the DOS at the Fermi energy decreases significantly.

In order to understand the changes we concentrate on the decomposition of the t_{2g} DOS into its symmetry components. Concerning the shapes of the partial d_{xz}/d_{yz} densities of states, there is a close similarity in the results for the rutile and the monoclinic structure. However, in the low temperature configuration both curves are subject to a considerable energetical upshift leaving the states almost unoccupied. The most striking deviations in the t_{2g} group of bands are due to the strong splitting of the partial $d_{x^2-y^2}$ DOS into two peaks. These findings for monoclinic VO_2 agree well with the band scheme proposed by J. B. Goodenough. Thus it is likely to attribute the behaviour of the $d_{x^2-y^2} = d_{\|}$ states to the V-V pairing along the rutile c-axis and the bonding-antibonding splitting arising from it. In addition, the energetical upshift of both the d_{xz} and d_{yz} bands (named π^* by J. B. Goodenough) is a consequence of lateral antiferroelectric-like displacements of the vanadium atoms and the resulting increase of the bonding between the V $3d$ and O $2p$ orbitals. For monoclinic VO_2 the two bands directly below the Fermi level separate from the other t_{2g} states. However, the split-off doublet still touches the next higher band at the A point, see figure 3.5. In fact, we are confronted with a non-vanishing density of states at the Fermi energy spoiling the experimental finding of an insulating ground state with an optical band gap of about 0.7 eV. Due to the partially filled conduction band at the C point the LDA calculation yields a semimetal-type behaviour for monoclinic VO_2. It is reasonable to attribute the failure in reproducing an insulating gap to the shortcomings of the local density approximation. Band gaps due to the bonding-antibonding splitting of hybridized bands appear to be poorly described in LDA calculations. Apart from the

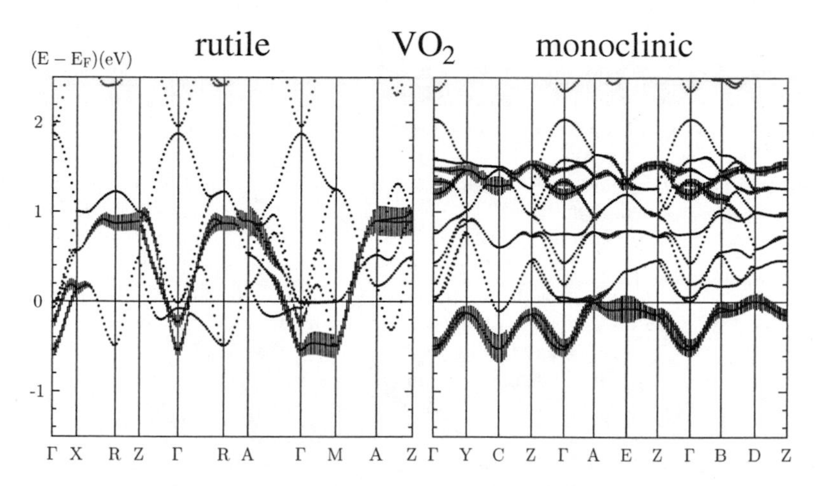

Figure 3.8: *Weighted electronic bands of rutile and monoclinic VO_2 depicted along selected symmetry lines in the first Brillouin zone of the simple tetragonal and simple monoclinic lattice, respectively. Regarding the definition of the given symmetry points compare figure 3.6. The width of the bars drawn for each band and k-point indicates the contribution due to the V $3d_{x^2-y^2}$ orbital located at site (0,0,0) relative to the local rotated reference frame.*

energetical positions of the $d_{\parallel}$ branches the agreement of calculated and measured band structures for both rutile and monoclinic VO_2 qualitatively is very good [44, 59].
Aiming at a deeper understanding of the mechanisms underlying the MIT of vanadium dioxide, weighted band structures are shown in figure 3.8. The length of the bars drawn for each band and **k**-point indicates the contribution from the V $3d_{x^2-y^2}$ orbital. In the rutile case one easily recognizes regions almost without any dispersion of the highlighted bands. They are connected by dispersing bands, particularly along high symmetry lines which comprise portions along the rutile c-axis (Γ-A, Γ-R, and Γ-Z). The meaning of the lines becomes clear from the tetragonal Brillouin zone in figure 3.6, if one bears in mind that the k_z-axis is parallel to c_{rut}. In contrast, the dispersion parallel to the rutile basal plane is suppressed to a large degree and the bands stay above or below the Fermi level. This is particularly true for the lines Γ-M, Z-A, Z-R, and R-A. Contributions due to the V $3d_{xz}$ or V $3d_{yz}$ orbital are covered by points without bars within the t_{2g} energy range of figure 3.8. This assignment is possible since neither O $2p$ nor V $3d$ e_g states give relevant contributions here. Apparently, the e_g^{π} bands are characterized by a rather isotropic dispersion compared to the $d_{x^2-y^2}$ states. As the hybridization between both types of bands is rather small, they are regarded as largely independent.
Considering the weighted band structure of monoclinic VO_2 reveals striking differences to the rutile case. One observes a strong splitting of the $d_{x^2-y^2}$ states into two narrow subbands at the lower and upper boundary of the t_{2g} region. This behaviour is consistent with the formation of a two peak structure in the partial $d_{x^2-y^2}$ DOS of the monoclinic phase,

see figure 3.5. In addition, the energetical upshift due to the zigzag-type antiferroelectric displacements of the vanadium atoms affects only the e_g^π bands. Being about 0.5 eV, this shift is matching the experimental data by S. Shin *et al.* [53]. With only minor exceptions the general topology of the e_g^π bands stays intact although they are depopulated to a large degree. Both the zigzag-type in-plane distortions and the vanadium dimerization induce antiferroelectric modes and thus contribute to the energetical upshift of the e_g^π states in the low temperature configuration. This fact was established by analyzing the M_2 phase of VO_2, in which each vanadium chain exhibits either zigzag-type displacements or V-V pairing [44]. Since both types of V $3d$ t_{2g} bands still hardly hybridize in the case of the M_1 structure, the changes at the phase transition presumably occur rather independently. To be more exact, bands of $d_{x^2-y^2}$ and e_g^π symmetry couple only by means of the common Fermi energy. As a consequence, the insulating state can be interpreted as arising from a Peierls-like instability of the $d_\parallel$ bands in an embedding background of the e_g^π electrons. Referring to studies on MoO_2 and NbO_2 [42,43] this mechanism appears to be typical for the class of the early transition metal oxides. Most notably, it gives rise to a unified explanation of the destabilization of the rutile structure in terms of an increased metal-metal bonding and hence clarifies the MITs in the d^1 compounds VO_2 and NbO_2.

3.3 Crystal structure of V_2O_3

As vanadium dioxide, the sesquioxide V_2O_3 has been attracting a lot of interest during the last decades. For that reason we include the compound in our considerations for the class of the vanadium oxides. In a first step, we again investigate the crystal structure – where we pay special attention to its relations to the structure of VO_2, which has already been discussed in great detail. Systematic studies have provided the temperature, pressure, and composition dependence of the $(V_{1-x}Cr_x)_2O_3$ phase diagram shown in figure 3.9 [61,62]. Stoichiometric V_2O_3 at ambient pressure undergoes an MIT at 168 K accompanied by a transformation from the high temperature corundum to the low temperature monoclinic structure. Simultaneously, an antiferromagnetic order develops (AFI phase). On doping by small amounts (a few percent) of chromium or aluminum an additional paramagnetic insulating configuration (PI phase) appears, which retains the full corundum symmetry of the paramagnetic metal (PM phase). As the metal-insulator boundary terminates at a critical point, a gradual crossover from the PM to the PI phase is observed when decreasing the temperature from well above the 500 K region. For each chromium concentration $0.005 \leq x_{Cr} \leq 0.018$ three phase transitions are encountered: PM to PI, PI to PM, and finally PM to AFI, where the last two are of first order [63]. For $x_{Cr} < 0.005$ the reentrant MIT disappears and for $x_{Ti} > 0.05$ the metallic phase extends over the whole temperature range. Hydrostatic pressure has almost the same effect as doping.
The corundum structure of metallic V_2O_3 consists of a trigonal lattice with space group $R\bar{3}c$ (D_{3d}^6). In the literature, usually a non-primitive hexagonal cell comprising six formula units is used, whereas the primitive trigonal cell contains two formula units. Structural refinements of V_2O_3 single crystals by P. D. Dernier [64] yield the hexagonal lattice constants $a_H = 4.9515$ Å and $c_H = 14.003$ Å. We find the vanadium atoms at the Wyckoff positions (12c): $\pm(0,0,z_V)$, $\pm(0,0,1/2+z_V)$, $\pm(-1/3,1/3,1/3)\pm(0,0,z_V)$, and $\pm(-1/3,1/3,1/3)\pm$

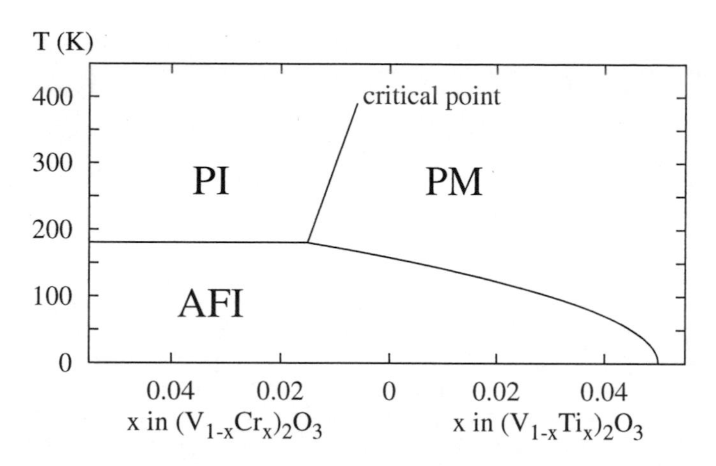

Figure 3.9: *Generalized phase diagram for the metal-insulator transition in V_2O_3 [61, 62]. The temperature dependence of the transition is depicted as a function of doping. Applying pressure instead of doping yields similar effects. Besides the paramagnetic metallic (PM) phase one observes paramagnetic insulating (PI) and antiferromagnetic insulating (AFI) phases. The line of the high temperature metal-insulator transition ends at a critical point.*

$(0, 0, 1/2+z_V)$. The oxygen atoms occupy the positions (18e): $\pm(x_O, 0, 1/4)$, $\pm(0, x_O, 1/4)$, $\pm(-x_O, -x_O, 1/4)$, $\pm(-1/3, 1/3, 1/3) \pm (x_O, 0, 1/4)$, $\pm(-1/3, 1/3, 1/3) \pm (0, x_O, 1/4)$, and $\pm(-1/3, 1/3, 1/3) \pm (-x_O, -x_O, 1/4)$. Here the positional parameters are given by $z_V = 0.34630$ and $x_O = 0.31164$ and the atomic coordinates in the hexagonal notation refer to the primitive translations

$$\mathbf{a}_H = \begin{pmatrix} 0 \\ -a_H \\ 0 \end{pmatrix}, \quad \mathbf{b}_H = \begin{pmatrix} \frac{\sqrt{3}}{2}a_H \\ \frac{1}{2}a_H \\ 0 \end{pmatrix}, \quad \mathbf{c}_H = \begin{pmatrix} 0 \\ 0 \\ c_H \end{pmatrix}. \tag{3.4}$$

Because there are three times as many atoms in the hexagonal than in the trigonal cell the use of the latter is favorable for calculations. We define the primitive translations of the trigonal lattice in the conventional way

$$\mathbf{a}_T = \begin{pmatrix} \frac{1}{2}a_T \\ -\frac{\sqrt{3}}{2}a_T \\ c_T \end{pmatrix}, \quad \mathbf{b}_T = \begin{pmatrix} \frac{1}{2}a_T \\ \frac{\sqrt{3}}{2}a_T \\ c_T \end{pmatrix}, \quad \mathbf{c}_T = \begin{pmatrix} -a_T \\ 0 \\ c_T \end{pmatrix}. \tag{3.5}$$

The trigonal lattice constants are connected to the hexagonal lattice constants via $a_H = \sqrt{3}a_T$ and $c_H = 3c_T$. After combining equations (3.4) and (3.5) it is possible to state the transformation matrix between the hexagonal and the trigonal lattice

$$\begin{pmatrix} \mathbf{a}_H \\ \mathbf{b}_H \\ \mathbf{c}_H \end{pmatrix} = \begin{pmatrix} 1 & -1 & 0 \\ 0 & 1 & -1 \\ 1 & 1 & 1 \end{pmatrix} \begin{pmatrix} \mathbf{a}_T \\ \mathbf{b}_T \\ \mathbf{c}_T \end{pmatrix} \tag{3.6}$$

Thus we are able to transform positions given in terms of the hexagonal primitive translations into their representations in the trigonal lattice

$$x\mathbf{a}_H + y\mathbf{b}_H + z\mathbf{c}_H = (x+z)\mathbf{a}_T + (-x+y+z)\mathbf{b}_T + (-y+z)\mathbf{c}_T \tag{3.7}$$

or equivalently

$$\begin{pmatrix} x \\ y \\ z \end{pmatrix}_T = \begin{pmatrix} 1 & 0 & 1 \\ -1 & 1 & 1 \\ 0 & -1 & 1 \end{pmatrix} \begin{pmatrix} x \\ y \\ z \end{pmatrix}_H . \tag{3.8}$$

Hence the vanadium atoms in the trigonal lattice are located at the Wyckoff positions (4c): $\pm(z_V^*, z_V^*, z_V^*)$ and $\pm(1/2 + z_V^*, 1/2 + z_V^*, 1/2 + z_V^*)$. Furthermore, the oxygen atoms take the positions (6e): $\pm(x_O^*, 1/2 - x_O^*, 1/4)$, $\pm(1/4, x_O^*, 1/2 - x_O^*)$, and $\pm(1/2 - x_O^*, 1/4, x_O^*)$. To complete the representation we state the positional parameters $z_V^* = z_V = 0.34630$, $x_O^* = x_O + 1/4 = 0.56164$ and the lattice constants $a_T = 2.85875$ Å, $c_T = 4.66767$Å.
Similar to the crystal structure of VO_2, the vanadium atoms in the sesquioxide are octahedrally coordinated by six oxygen atoms. Qualitatively, the oxygen network is likewise equivalent. In particular, in the c-direction of the non-primitive hexagonal unit cell the octahedra are linked via faces. Hence we identify the c_{hex}-axis with the a_{prut}-axis. Since only two thirds of the octahedra are filled, no continuous chains of face-sharing VO_6 units arise, but two filled octahedra are always followed by one empty. Neighbouring chains are shifted along c_{hex} in a way that hexagonal structures of vanadium atoms are generated. The latter observation forms the basis of describing a corundum structure in terms of a hexagonal cell. Figure 3.10 schematically illustrates the atomic arrangement of the vanadium sites in V_2O_3. Due to the trigonal symmetry every vanadium atom exhibits three identical in-plane V-V distances of 2.87 Å. Furthermore, there is one shorter distance of 2.70 Å to the adjacent vanadium site along c_{hex}. Within each VO_6 unit one finds three V-O bond lengths of 1.97 Å and three other of 2.05 Å. As the V-V interaction through the octahedral faces (along c_{hex}) is almost unhindered, the hexagonal in-plane vanadium structures deviate from being planar, which can be understood from simple electrostatic considerations. At first glance one would expect the alternating V-V distances along c_{hex} to fulfill a ratio of 2:1. In contrast to this assumption the bond lengths are reported to be 2.70 Å and 4.30 Å. Mainly due to ionic interactions the vanadium pairs are subject to an anti-dimerization and hence to an increase of their mutual distance. In comparison to the expected value of 2.33 Å the measured bond length in the vanadium pairs deviates by more than 15%.
At approximately 168 K corundum V_2O_3 distorts into a monoclinic structure. Structural refinements of the low temperature phase show an increase of the nearest neighbour V-V distances [65]. The bond lengths across the shared octahedral faces (along a_{prut}) increase from 2.70 Å to 2.75 Å. Moreover, the threefold degeneracy of the in-plane V-V distance is lifted, which gives rise to a remarkably increased (2.99 Å) bond across an octahedral edge. We identify the direction of this bond as the c_{prut}-axis. The symmetry breaking at the PM to AFI transition yields two crystallographically inequivalent oxygen sites. An oxygen octahedron comprises four atoms of the first and two atoms of the second kind. The distances within the VO_6 unit vary in the region from 1.95 Å to 2.11 Å, whereas the average length of the six bonds remains essentially constant compared to the PM phase.

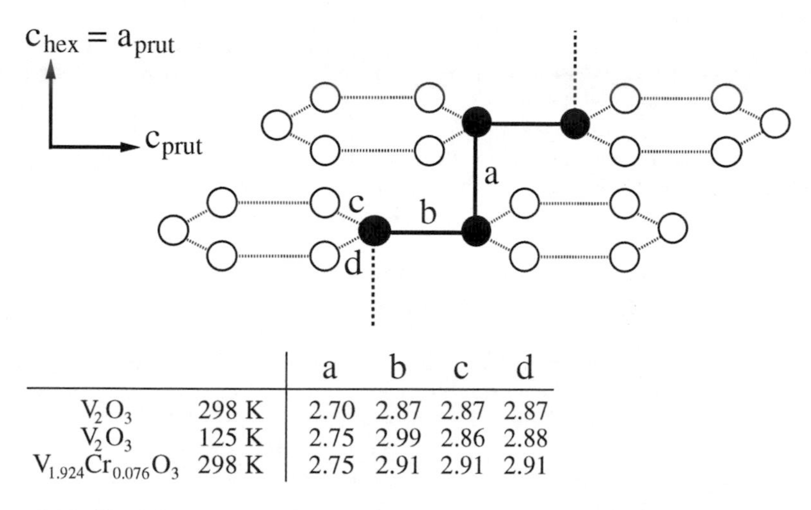

		a	b	c	d
V_2O_3	298 K	2.70	2.87	2.87	2.87
V_2O_3	125 K	2.75	2.99	2.86	2.88
$V_{1.924}Cr_{0.076}O_3$	298 K	2.75	2.91	2.91	2.91

Figure 3.10: *Crystal structure of V_2O_3. For simplicity the schematic view does not include the oxygen sublattice, which is an octahedral network qualitatively equal to the case of VO_2. The vanadium atoms once more occupy octahedral sites but now form hexagonal structures perpendicular to the hexagonal c_{hex}-axis. Hence vanadium pairs arise both along $a_{prut}=c_{hex}$ and c_{prut}. The table records measured V-V distances (in Å) for the paramagnetic metallic, the antiferromagnetic insulating, and the paramagnetic insulating phase of V_2O_3 [64, 65].*

The monoclinic cell of AFI V_2O_3 is body-centered with space group $I2/a$ (C_{2h}^6). Starting with the conventional representation of the simple monoclinic unit cell

$$\mathbf{a}_M = \begin{pmatrix} 0 \\ 0 \\ -a_M \end{pmatrix}, \quad \mathbf{b}_M = \begin{pmatrix} -b_M \\ 0 \\ 0 \end{pmatrix}, \quad \mathbf{c}_M = \begin{pmatrix} 0 \\ c_M \sin\beta_M \\ -c_M \cos\beta_M \end{pmatrix} \tag{3.9}$$

we apply the transformations $\mathbf{a}_{BM} = \frac{1}{2}(\mathbf{a}_M + \mathbf{b}_M + \mathbf{c}_M)$, $\mathbf{b}_{BM} = \mathbf{b}_M$, and $\mathbf{c}_{BM} = \mathbf{c}_M$ [66] to obtain the primitive translations of the body-centered monoclinic lattice. P. D. Dernier and M. Marezio [65] reported the values $a_M = 7.255\,\text{Å}$, $b_M = 5.002\,\text{Å}$, $c_M = 5.548\,\text{Å}$, and $\beta_M = 96.75°$ for the lattice constants and the monoclinic angle. The body-centered unit cell comprises four formula units. Both the vanadium atoms and the first kind of oxygen atoms occupy the Wyckoff positions (8f): $\pm(x, y, z)$, $\pm(1/2 + x, -y, z)$, $\pm(1/2 + x, 1/2 + y, 1/2 + z)$, and $\pm(x, 1/2 - y, 1/2 + z)$. The parameters $(x, y, z)_V = (0.3438, 0.0008, 0.2991)$ and $(x, y, z)_{O1} = (0.407, 0.845, 0.652)$ fix the atomic positions in the unit cell. In addition, the second oxygen class is found at the Wyckoff positions (4e) with parameter $y_{O1} = 0.191$: $\pm(1/4, y_{O1}, 0)$ and $\pm(3/4, 1/2 + y_{O1}, 1/2)$.
In contrast to the symmetry breaking at the PM-AFI transition of V_2O_3 the corundum structure is conserved when going from the PM to the PI phase [64]. The hexagonal lattice constants $a_H = 4.9985$, $c_H = 13.912$ and the positional parameters $z_V = 0.34870$, $x_O =$

0.30745 change only slightly. Nevertheless, the V-V bond lengths through octahedral faces and through octahedral edges increase to 2.75 Å and 2.91 Å, respectively. A comparison of the V-V bond lengths in the three phases of vanadium sesquioxide is included in figure 3.10. Using hard and soft x-ray absorption techniques a comparative investigation of the PI and AFI phases of aluminum doped V_2O_3 revealed an identical local symmetry of the vanadium sites [67]. On a local scale the structural distortions thus seem to be the same in both insulating phases.

3.4 Electronic Properties of V_2O_3

Vanadium sesquioxid has been analyzed extensively as the canonical Mott–Hubbard system [47, 48]. The description of V_2O_3 in terms of the one-band Hubbard model is based on a level scheme for the electronic structure presented by C. Castellani *et al.* [68]. Very similar to the case of VO_2 the crystal field generated by the oxygen octahedra splits the V 3d orbitals into lower t_{2g} and higher e_g^σ states. The former split further into a_{1g} and e_g^π levels due to the trigonal symmetry of the lattice. Electronic interactions between nearest vanadium neighbours along c_{hex} yield bonding and antibonding molecular orbitals of the a_{1g} states. While the bonding orbital is completely occupied, the antibonding states shift energetically above the e_g^π levels. This leaves one electron per vanadium site in the twofold degenerate e_g^π states, making the system susceptible to degeneracy lifting distortions such as the Jahn–Teller effect and orbital ordering. Moreover, the non-degenerate a_{1g} orbital points parallel to c_{hex} and gives rise to a one-dimensional band due to covalent bonding. The total spin $S = 1/2$ of the single e_g^π electron suggests using the half filled one-band Hubbard model as the simplest possible model describing vanadium sesquioxide.
The temperature dependent MIT in (undoped) V_2O_3 between the high temperature PM and the low temperature AFI phase was experimentally analyzed by photoelectron spectroscopy. As reported by S. Shin *et al.* [53], the spectra display the O 2p band located in the energy range from approximately -10 eV to -4 eV, whereas the V 3d band is settled within about 3 eV below the Fermi level. Except for slight modifications of the bandwidth no drastic change is found in the V 3d band structure below the transition. However, the DOS at the Fermi energy is rather small – even in the metallic phase. A high resolution photoemission study by S. Shin *et al.* [69] revealed a band gap of 0.2 eV for the insulating phase. In the metallic case finite spectral weight appears at the Fermi energy but the lower Hubbard band persists almost without change in its line shape, indicating that electronic correlations remain substantial even in the metallic phase. The high temperature MIT in chromium doped samples of V_2O_3 between the PI and the PM phase was investigated by K. E. Smith and V. E. Henrich [70] using photoelectron spectroscopy. In the insulating regime at room temperature they found a low emission intensity at the Fermi level, which increases after cooling into the metallic state. The high temperature PI spectra reveal a larger V 3d bandwidth than the low temperature AFI spectra, which may be due to the thermal broadening and the absence of a magnetic order. According to oxygen K-edge x-rax absorption spectroscopy additional unoccupied V 3d bands extend from the Fermi level to about 6 eV [71]. One observes broad maxima near 1 eV and 3 eV.
Band structure calculations confirmed the electronic level scheme proposed by C. Castel-

lani *et al.* with the t_{2g} orbitals located in the vicinity of the Fermi energy [72]. Recently, the model has been called into question due to results of polarized x-ray absorption spectroscopy, which call for a vanadium $S = 1$ spin state [73]. Furthermore, the first excited states miss the expected pure e_g^π symmetry but include remarkable a_{1g} admixtures. This requires an explanation beyond the simplicity of the pure one-band Hubbard model or of models projecting out the a_{1g} orbital by means of simple dimerization. LDA+U calculations succeeded in explaining the electronic and magnetic properties of the AFI phase, in particular the peculiar antiferromagnetic order [74]. While each vanadium spin is parallel to the adjacent spin along c_{hex}, there are two antiparallel spins plus one parallel spin at the nearest in-plane vanadium sites. Enforcing spin degeneracy even for the AFI phase the influence of the crystal structure on the electronic properties was found to be rather small. Introducing the antiferromagnetic order yields an optical band gap of 0.6 eV. Moreover, the LDA+U results point at a spin $S = 1$ model without orbital degeneracy. In contrast, by means of model calculations F. Mila *et al.* [75] proposed a spin state $S = 1$ and orbital degeneracy in the AFI phase. Starting with the assumption of strong covalent bonding in the vanadium pairs along c_{hex} these authors supposed the intersite a_{1g} hopping matrix elements to dominate. However, a recent comparison of the hopping processes in V_2O_3 revealed hopping integrals between second, third, and fourth nearest vanadium neighbours being equally important for the shape of the a_{1g} band [76]. Evidence for orbital ordering seems to arise from magnetic neutron scattering [77]. The magnetic fluctuations in the AFI phase are fundamentally different from those observed in the PM and the PI phase. Due to the orbital ordering the AFI phase would be distinguished from the PI phase thus preventing a unified description of the phase transitions in vanadium sesquioxide. LDA band structure calculations performed for all three phases of V_2O_3 show only minor response of the electronic structure to the crystal parameter changes. Nevertheless, a slight narrowing of the characteristic a_{1g} bands in the insulating phases formed the basis for a successful description of the PM-PI transition by a combination of LDA calculations with the dynamical mean field theory (DMFT) [78, 79]. In the same manner as the mentioned LDA+U results, this demonstrates the strong influence of correlations on the electronic structures of the insulating phases. Confirming predictions of the LDA+DMFT approach, a prominent quasiparticle peak was observed in the photoemission spectrum of metallic V_2O_3 [80]. However, despite an enormous number of papers the electronic properties and metal-insulator transitions of vanadium sesquioxide are still controversial topics.

To understand the results for the vanadium Magnéli phases presented in the next chapter we now review findings of LDA band structure calculations for V_2O_3. Figure 3.11 displays the bands calculated from the crystallographic data of the PM corundum structure and the AFI monoclinic configuration. Because we aim at an investigation of the dependences connecting the crystal structure and electronic behaviour, spin degeneracy is enforced in the AFI calculation. The electronic bands are depicted along selected symmetry lines in the first Brillouin zone of the trigonal and the simple monoclinic lattice, respectively. A definition of the high symmetry points is given in figures 3.6 and 3.12.

The common structural properties of VO_2 and V_2O_3 imply some agreement of the gross features of the band structures, see figures 3.5 and 3.11. As for the dioxide, we observe three groups of bands for both corundum and monoclinic V_2O_3. In the corundum case they are settled in the energy regions from -8.8 eV to -3.9 eV, from -1.1 eV to 1.5 eV,

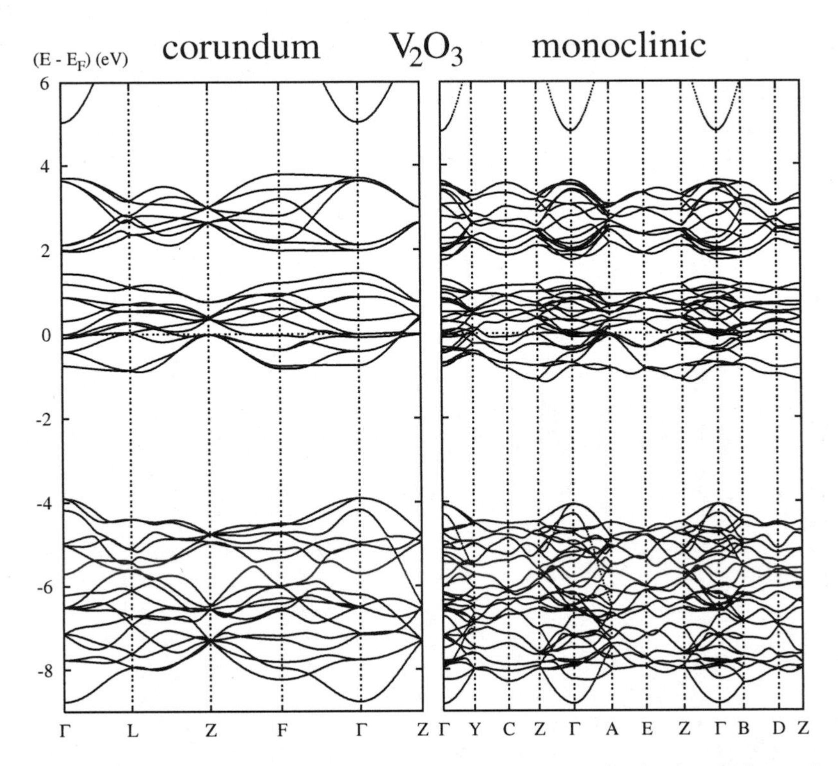

Figure 3.11: *Electronic bands of corundum and monoclinic V_2O_3 displayed along selected symmetry lines in the first Brillouin zone of the trigonal and the simple monoclinic lattice, respectively. Regarding the definition of the given symmetry points see figures 3.6 and 3.12.*

and from 1.9 eV to 3.8 eV. Only minor modifications of the energetical positions appear in the monoclinic case, except for small shifts of the t_{2g} and e_g^σ states. While the former slightly narrow and occupy the energy range between −1.2 eV and 1.3 eV, the latter move about 0.2 eV towards the Fermi level. The energetical downshift of all bands, compared to the findings for VO_2, is due to the modified electron count in the d^2 system V_2O_3. In the case of rutile VO_2 with two formula units per unit cell the observed groups of states contain 12, 6, and 4 bands. Two times five V 3d orbitals give rise to ten vanadium dominated bands and four times three O 2p orbitals explain the remaining twelve bands with oxygen predominance. Using the same simple arguments we expect 18, 12, and 8 bands for corundum V_2O_3, which likewise has two formula units per unit cell. From figure 3.11 this assumption is confirmed. Due to the four formula units per monoclinic cell the low temperature findings comprise twice as many bands. So far, the molecular orbital picture derived for VO_2 applies to vanadium sesquioxide as well.
In the next step we turn to the V 3d and O 2p densities of states in the upper section of

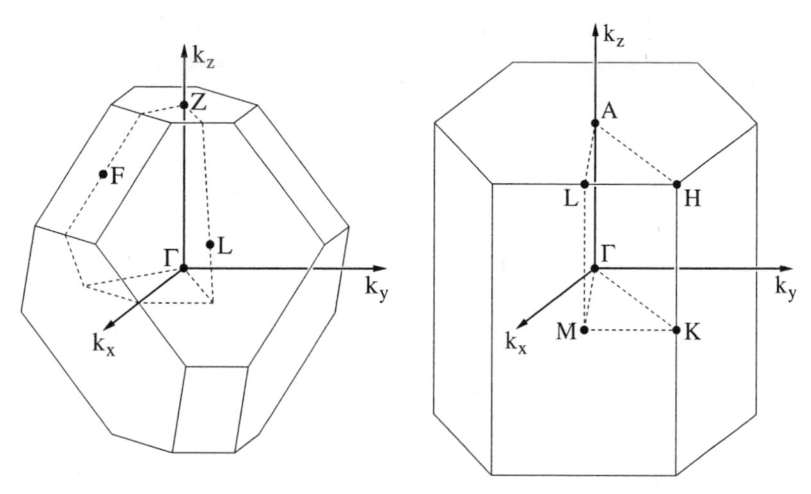

Figure 3.12: *First Brillouin zones of the trigonal (or rhombohedral; left) and the hexagonal (right) lattice. Dashed lines mark the irreducible wedge and labels denote symmetry points.*

figure 3.13. As is typical for transition metal chalcogenides with octahedral coordination we find three distinct structures in the DOS reflecting the three groups of bands previously discussed. While the lowest group mainly traces back to O $2p$ orbitals, the groups at and above the Fermi energy primarily are due to V $3d$ contributions. Hybridization between vanadium and oxygen leads to significant admixtures of the V $3d$ and O $2p$ states in the energy intervals dominated by the respective other states. Due to the σ-type overlap the mixing is increased in the σ and σ^*-areas. Generally speaking, the magnitude of the p-d hybridization is of the same order as found for the dioxide. Obviously, the t_{2g} occupation increases on passing from the d^1 to the d^2 system.
Owing to their relevance in the context of the phase transitions we study the t_{2g} group of states separated into its symmetry components in the lower row of figure 3.13. First, one observes noticeable contributions of the t_{2g} orbitals in the e_g^σ energy range above 1.9 eV for the corundum and above 1.7 eV for the monoclinic structure. They are mainly caused by the anti-dimerization of the V-V pairs along the c_{hex}-axis, which shifts the vanadium atoms away from the centers of the surrounding oxygen octahedra. Hence the octahedral coordination of these vanadium sites is perturbed and the separation in t_{2g} and e_g^σ states is no longer ideal. Second, one observes only minor differences in the results for the three symmetry components, particularly in the corundum case. Comparing the corresponding VO_2 DOS (second row of figure 3.7) this behaviour clearly distinguishes the compounds. The specific answer of the electronic properties to modifications of the crystal structure accompanying the MIT of VO_2 paved way for interpreting the transition as an embedded Peierls instability. A similar procedure seems to fail in the case of the sesquioxide because the structural modifications influence the outcome of the LDA calculation only to a minor degree. The latter particularly is true for the PI structure, for which the respective results

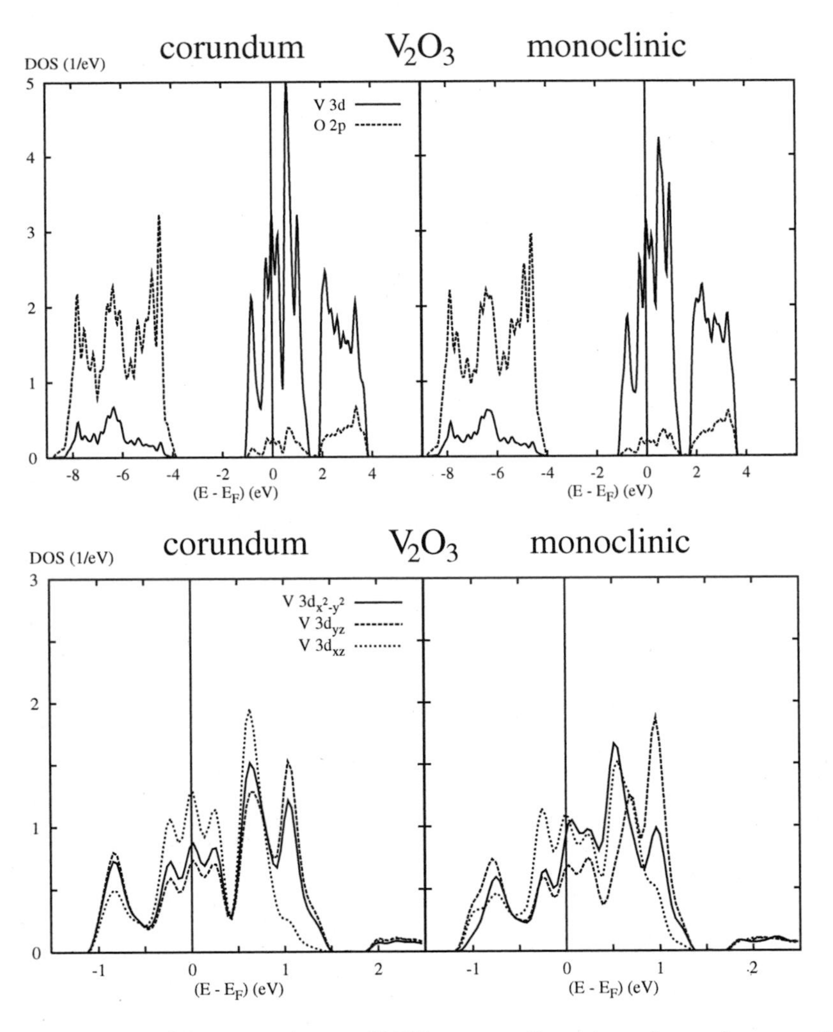

Figure 3.13: *Partial densities of states (DOS) per vanadium atom of corundum as well as monoclinic V_2O_3. The first row presents a comparison of the partial V 3d and O 2p DOS resulting from the crystal structures of paramagnetic metallic (left) and antiferromagnetic insulating (where spin degeneracy was enforced; right) V_2O_3. In the second row the partial V 3d t_{2g} DOS is shown in detail. Here the orbitals refer to the local rotated reference frame.*

are not shown. However, a way out of this situation later will arise from the investigation of the vanadium Magnéli phases. That the three V_2O_3 t_{2g} densities of states resemble one

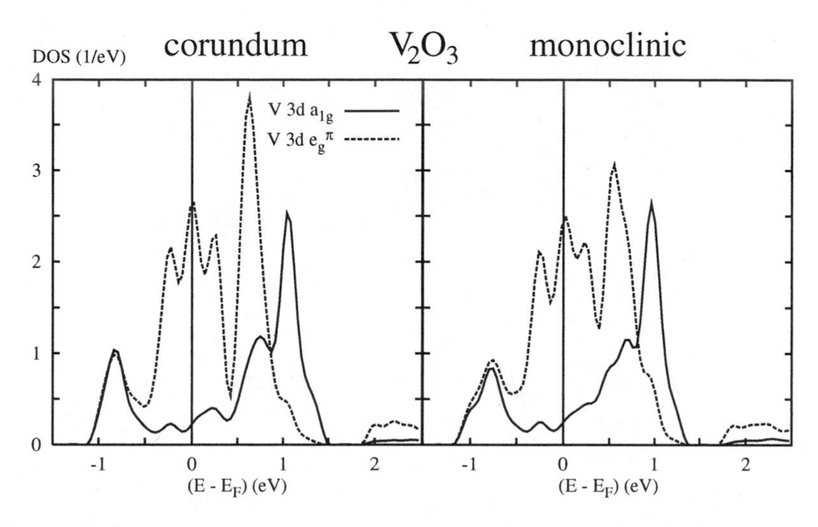

Figure 3.14: *Partial V 3d a_{1g} and V 3d e_g^π densities of states (DOS) per vanadium atom of corundum and monoclinic V_2O_3; alternative representation of the results in figure 3.13.*

another to a high degree is just a consequence of the trigonal symmetry of the corundum structure, which is only slightly perturbed below the MIT. Obviously, there is no sign of an emerging energy gap in the spin degenerate LDA calculation for the AFI phase. Since the same is reported for the PI phase, the lacking of an insulating gap presumably is not due to the enforced spin degeneracy but traces back to electronic correlations, which the local density approximation does not adequately account for.
The presentation of the V_2O_3 t_{2g} data in figure 3.13 refers to the analogous local rotated reference frame as applied in figure 3.7, allowing for a straightforward comparison of the results. However, due to the trigonal symmetry an alternative description may be appropriate. According to C. Castellani *et al.*, who proposed the V $3d$ t_{2g} orbitals to split up into a_{1g} and e_g^π states, we split the t_{2g} DOS into its symmetry components in figure 3.14. Actually this kind of representation offers the advantage of indicating a one-dimensional band growing out of the a_{1g} states. Referring to the molecular orbital picture, we identify the distinct peaks at approximately $-0.8\,\text{eV}$ and $1.0\,\text{eV}$ as the bonding and antibonding states due to the in-pair interaction of nearest vanadium neighbours in the c_{hex}-direction. Confirming the expectation, the bonding states are completely filled and the antibonding orbitals remain empty. Surprisingly, the weights of the bonding and the antibonding peak are far from equal but reveal a ratio of 1:3. This observation is hard to explain from the molecular orbital point of view, casting doubts on the one-dimensional nature of the a_{1g} component. Hence we address this question again in a later chapter. Due to the reduced weight of the bonding a_{1g} peak more than one electron occupies the twofold degenerate e_g^π orbitals. In the vicinity of the Fermi energy the majority of the spectral weight traces back to a characteristical three peak structure due to the latter orbitals. Further e_g^π states

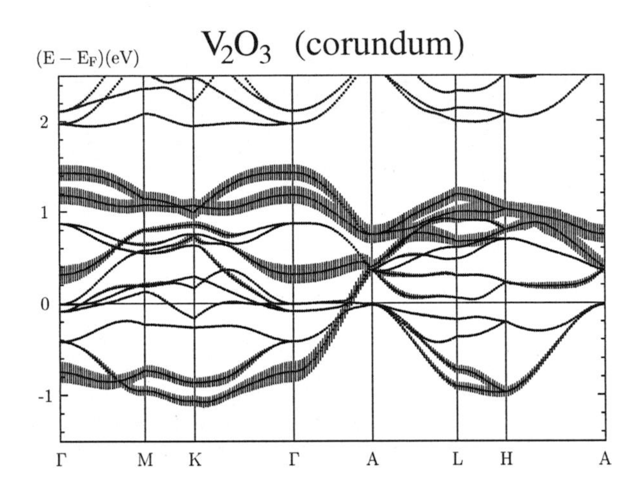

Figure 3.15: *Weighted electronic bands of corundum V_2O_3 shown along selected symmetry lines in the first Brillouin zone of the hexagonal lattice, compare figure 3.12. The width of the bars included for each band and k-point indicates the contribution due to the V 3d a_{1g} orbital of the atom located at site (z_V^*,z_V^*,z_V^*) relative to the local rotated reference frame.*

are found at about 0.7 eV in the high temperature phase and slightly closer to the Fermi energy in the monoclinic phase. The shape of the a_{1g} DOS hardly changes at the MIT. Only a little narrowing of the whole t_{2g} group by about 0.1 eV is observed.
To better understand the origin and the shape of the a_{1g} DOS a weighted band structure for the corundum modification is given in figure 3.15. As a consequence of the structural similarities, the results for the low temperature monoclinic phase are closely related (not depicted). The length of the bars added to the bands represents the magnitude of the a_{1g} contributions to the particular states. In contrast to the band structure in figure 3.11 we do not use the primitive trigonal lattice but refer to the non-primitive hexagonal representation, see equation (3.4). The first Brillouin zone of the hexagonal lattice is illustrated in figure 3.12. Due to the discussed hexagonal arrangement of the vanadium sites in V_2O_3, the high symmetry lines of the hexagonal Brillouin zone are correlated with directions of interest in real space. In particular, the reciprocal k_z-axis is parallel to the c_{hex}-axis along which the a_{1g} orbitals are oriented. In general, the a_{1g}-like bands in figure 3.15 surround the remaining e_g^π-like bands, which are depicted by points without bars in the t_{2g} energy range. While the dispersion of the e_g^π bands is rather isotropic, the a_{1g} dominated states reveal an increased dispersion along the line Γ-A, i.e. in the c_{hex}-direction. The fact that the a_{1g}-like bands cross the Fermi energy (Γ-A) contradicts an ordinary interpretation in terms of bonding-antibonding split one-dimensional states. Finally, the bands located at 1.2 eV and 1.4 eV at the Γ point have a rather undisturbed a_{1g} character. They seem well separated from the e_g^π-like bands, which is not the case for the lower a_{1g} branch.

Chapter 4

Vanadium Magnéli Phases

The vanadium oxides comprise compounds with several formal vanadium valency stages, which reach from two in VO, three in V_2O_3, and four in VO_2 to five in V_2O_5. In addition to these configurations mixed valent compounds can be synthesized. Amongst the latter the so-called Magnéli phases, defined by the general stoichiometric formula

$$V_nO_{2n-1} = V_2O_3 + (n-2)VO_2 \quad \text{where} \quad 3 \leq n \leq 9\,, \tag{4.1}$$

are of special interest as they give rise to a homologous series of compounds with closely related crystal structures. This kind of homologous series has been reported for the first time by A. Magnéli in the case of the molybdenum oxides [82]. Today Magnéli series are known from the vanadium, titanium, niobium, and tungsten oxides. The first structural characterization of vanadium Magnéli phases by means of x-ray investigations traces back to S. Andersson and L. Jahnberg [83] in 1963. As equation (4.1) indicates, the Magnéli phases take an intermediary position between V_2O_3 and VO_2, thus between the valency stages three and four. In addition to the chemical relationsship, the crystal structures of the Magnéli phases actually consist of rutile and corundum-type blocks and consequently show structural affinity to both the dioxide and the sesquioxide. In the next section we study the structural relations between these materials in detail. We will thereby discover the potential of gradually transferring the crystal structure of VO_2 into that of V_2O_3 by making use of the Magnéli phases. Finally, this observation paves the way for a unifying picture of the whole Magnéli series including its end members VO_2 and V_2O_3.
As a function of temperature each Magnéli phase undergoes an MIT, except for V_7O_{13}, which stays metallic at all temperatures. These transitions are of first order, reduce the conductivity by several orders of magnitude, and are accompanied by distinct structural transformations. In table 4.1 transition temperatures from electrical resistivity measurements performed by S. Kachi *et al.* [84] are summarized. Moreover, this table compares the formal V $3d$ charges of the Magnéli compounds, varying from two in V_2O_3 to one in VO_2. In an ionic picture each vanadium atom contributes three $3d$ and two $4s$ electrons, whereas every oxygen atom accepts two additional $2p$ electrons. Considering V_3O_5, for example, this electron transfer yields a formal vanadium charge of $(3 \cdot 5 - 5 \cdot 2)/3 = 5/3$. As the Magnéli phases are characterized by vanadium atoms in mixed valent states, their electronic properties were expected to be influenced by charge order. Thus they are most

Compound V_nO_{2n-1}	Parameter n	Formal V $3d$ charge	MIT temperature
V_2O_3	2	2	168 K
V_3O_5	3	$5/3 \approx 1.67$	430 K
V_4O_7	4	$6/4 \approx 1.50$	250 K
V_5O_9	5	$7/5 \approx 1.40$	135 K
V_6O_{11}	6	$8/6 \approx 1.33$	170 K
V_7O_{13}	7	$9/7 \approx 1.29$	metallic
V_8O_{15}	8	$10/8 \approx 1.25$	70 K
V_9O_{17}	9	$11/9 \approx 1.22$	—
VO_2	∞	1	340 K

Table 4.1: *Formal V 3d charges as well as transition temperatures in the series V_nO_{2n-1}.*

interesting from both the experimental and theoretical point of view. Nevertheless, only few investigations for this material class are reported in the literature, which we summarize in the following.
Figure 4.2 shows the behaviour of the MIT temperatures listed in table 4.1 as a function of the vanadium-oxygen ratio. We recognize a broad minimum centered at the ratio corresponding to V_7O_{13}. Interestingly, the MIT of the vanadium Magnéli phases is coupled to an anomaly in the magnetic susceptibility, which closely resembles the characteristics of a transition from a paramagnetic to an antiferromagnetic state [84]. Nonetheless, this anomaly is not accompanied by the development of magnetic order. Yet, all the Magnéli phases enter an antiferromagnetic groundstate at sufficiently low temperatures. In some sense the anomaly thus may be interpreted as a precursor effect of these transitions. The behaviour of the respective Néel temperatures as a function of the vanadium oxygen ratio is also given in figure 4.2. Here we find a maximum in the case of V_2O_3 and a cumulative decrease on approaching VO_2, where an antiferromagnetic ordering no longer is observed. V_7O_{13} again gives rise to the only deviation from the trend. One possible explanation for the decreasing magnetic transition temperatures is the continuous reduction of V_2O_3-like regions in the crystal structure. However, this kind of mechanism has not been confirmed. The MIT and the magnetic ordering are coupled only in the case of the sesquioxide. For the other Magnéli phases they appear at different temperatures and are therefore likely to trace back to different mechanisms. Since we are mainly interested in the metal-insulator transitions, the electronic structure calculations presented subsequently do not take into account the low temperature antiferromagnetic order. Actually, it would not be possible to include the magnetism since the detailed magnetic structures of the ordered phases are not known.
Specific heat measurements above the magnetic ordering temperature, carried out by G. D. Khattak *et al.* [85] for all the Magnéli phases, reveal larger values than expected from pure lattice contributions. Together with the smallness of the magnetic entropy increase at the respective Néel temperature these authors argued for further magnetic disordering taking place above the phase transition. In order to explain these findings they proposed a model of linear antiferromagnetic chains. However, due to low temperature investigations of the magnetic susceptibility S. Nagata *et al.* [86] were able to rule out this model.

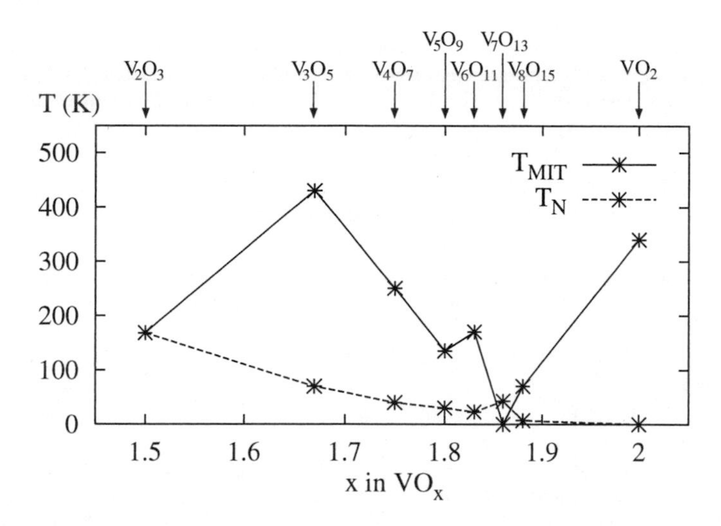

Figure 4.2: *Comparison of the MIT temperatures and the magnetic ordering temperatures in the Magnéli series. Besides* V_7O_{13} *all members of this series undergo an MIT. Except for* VO_2 *each compound develops an antiferromagnetic order at low temperatures. In the case of* V_2O_3 *the MIT and the magnetic ordering coincide at a temperature of circa 168 K.*

Furthermore, they suggested interpreting the insulating V_8O_{15} modification as a charge-density wave state. A. C. Gossard *et al.* [87] speculated about a charge-density wave in V_4O_7. From the comparison of the V_4O_7 and V_8O_{15} magnetic susceptibilities S. Nagata *et al.* [86] concluded the charge-density wave to be less complete in the former material. They attributed small magnetic moments at the vanadium sites, present in the insulating phases of both V_4O_7 and V_6O_{11}, to singlet pairing effects.
M. Marezio *et al.* [88] concluded from their single crystal x-ray investigations that charge localization is present in the insulating phase of V_4O_7. They reported chains of V^{3+} and V^{4+} ions running parallel to the c_{prut}-axis, whereas the vanadium valences are disordered in the metallic phase. This kind of crystallographic evidence for charge localization arises from a comparison of the measured V-O distances in V_4O_7 with estimated values referring to an ionic picture. VO_6 units with large and small average V-O distances are assigned to V^{3+} and V^{4+} states, respectively. However, due to the covalent portion of the interatomic bonding the calculated vanadium charges might be poor. Specific heat measurements by B. F. Griffing *et al.* [89, 90] implied that the metallic states of V_4O_7 and V_7O_{13} are characterized by the coexistence of itinerant and essentially localized V $3d$ electrons. Nuclear magnetic resonance experiments confirmed these results [87, 91]. The latter data revealed two inequivalent vanadium sites, but contradicted $S = 1/2$ and $S = 1$ spin states, which are appropriate to d^1 and d^2 ionic configurations, respectively. Hence the authors proposed a modification of the local-moment picture with narrow strongly correlated V $3d$ bands in the metallic state. Furthermore, in the insulating phase all Magnéli compounds

show indications of increased charge localization combined with singlet pairing, except for V_3O_5. Therefore the MITs were proposed to be combinations of arising magnetic short range order and charge localization. In particular, the charge localization would split the subbands associated with the inequivalent metal atoms and therefore might pave the way for a Mott–Hubbard transition.
While the MIT temperature of VO_2 continuously increases on the application of pressure, V_2O_3 reveals the opposite trend, see the generalized phase diagrams given in figures 3.3 and 3.9. The Magnéli phases themselves clearly reflect the behavior of the sesquioxide as their transition temperature decreases with growing hydrostatic pressure [33]. Moreover, P. C. Canfield *et al.* [92] reported on unifying aspects emerging in the Magnéli series on the application of pressure. They observed for both V_8O_{15} and $(V_{0.98}Cr_{0.02})_8O_{15}$ the phase transitions PM-AFM followed by AFM-PI. Hence the PM-AFM transition of V_7O_{13} at atmospheric pressure should not be regarded anomalous for the Magnéli series. While the PM-AFM transition appears in V_8O_{15} only above 9 kbar, it is found at all pressures for the chromium doped compound. The magnetic transition exhibits the same pressure dependence in all three cases ($-0.75\,\mathrm{K/kbar}$). Since the MIT temperature decreases at a larger rate the AFM modification is stabilized for sufficiently high pressures. In other Magnéli phases the PM-AFM transition is not observed up to roughly 20 kbar. Nevertheless, the authors assume different pressure dependences of the MIT and the antiferromagnetic ordering as a general feature, common to all Magnéli phases. Consequently, it is reasonable to suppose that both effects evolve independently. The latter strictly contradicts the findings for the sesquioxide thus implying a different mechanism of the MIT.
A comprehensive treatment of experimental and theoretical results for the vanadium oxides was given by W. Brückner *et al.* [33]. Summarizing, an understanding of the mechanisms underlying the MITs in the vanadium Magnéli class is not in reach. Proposed scenarios include Peierls-type mechanisms and charge localization, possibly combined with Mott–Hubbard schemes.

4.1 Representation of the Crystal Structures

Aiming at a profound analysis and discussion of the electronic structures of the vanadium Magnéli phases in the following sections, we first have to understand in detail the crystal structures, which is the goal of the subsequent considerations. A precise structural analysis is a necessary prerequisite in order to relate the electronic changes at the MITs of the Magnéli phases to the structural transformations occuring simultaneously. In particular, we will elaborate a new and unifying point of view of the atomic arrangements underlying all members of the series. An exciting advantage of this representation is its applicability to the crystal structures of vanadium dioxide and sesquioxide as well. Thus a comprehensive understanding of the whole Magnéli series V_nO_{2n-1} including its end members VO_2 ($n \to \infty$) and V_2O_3 ($n = 2$) is eventually achieved.
As already mentioned, and as is obvious from the stoichiometric relation in equation (4.1), the crystal structures of the Magnéli phases are usually viewed as rutile-type slabs separated by shear planes with a corundum-type atomic arrangement [83]. The rutile slabs extend infinitely in two dimensions and have a characteristic finite width corresponding

to n VO_6 octahedra in the case of the compound V_nO_{2n-1}. Since octahedra at a slab-surface share faces with octahedra from a neighbouring slab, the atomic arrangement at the boundary is closely related to the corundum structure. The rutile crystal structure is therefore disturbed and adjacent slabs are mutually out of phase, thus giving rise to the denotation shear plane.

Precise investigation of the changes of the local atomic coordination on passing through the Magnéli series is rather complicated using the ordinary representation of the crystal structures. Hence it is advantageous to describe the structures in a different manner. For that purpose we start with the oxygen sublattice, which turns out to be most similar for all the compounds. As demonstrated for VO_2 and V_2O_3 the sublattice can be understood as a regular space filling network of neighbouring oxygen octahedra. The octahedra are mutually connected via edges in the c_{prut}-direction and via faces along the perpendicular axes a_{prut} and b_{prut}. In order to illustrate this geometrical arrangement schematical projections of sandwich-like O-V-O slabs cut out of the crystal structures of the dioxide and the sesquioxide are depicted in figure 4.3. In each case the projection is parallel to a_{prut}. A sandwich-like slab consists of a vanadium layer confined by oxygen layers at both ends. As a consequence of the projection the oxygen octahedra appear as hexagonal structures and the octahedral network becomes a two-dimensional hexagonal network.

We clearly observe the infinite vanadium chains of VO_2 running along c_{prut} and the hexagonal configuration of the vanadium atoms in V_2O_3. Actually, the basic structural features of the dioxide and the sesquioxide are captured completely in this graphical representation. Adjacent slabs yield the same projection as shown except for a possible translation perpendicular to a_{prut}. To be more specific, in adjacent VO_2 slabs filled and empty oxygen octahedra, i.e. filled and empty oxygen hexagons, are interchanged. In neighbouring V_2O_3 slabs the vanadium sites are shifted by one hexagon along the c_{prut}-direction. To sum up, provided that the oxygen networks of VO_2 and V_2O_3 are exactly the same, the only difference between the compounds arises from a different vanadium sublattice. The configuration of the latter is optimally described in terms of the presented slab-type projection. In particular, the infinite vanadium chains belonging to the dioxide are contrasted with finite chains of length 2 in the case of the sesquioxide. Such a description of the V_2O_3 structure in terms of 2-chains along c_{prut} instead of hexagonal arrangements is preferable for discussion. In contrast to our assumption, one observes of course differences between the oxygen sublattices of VO_2 and V_2O_3. However, they do not affect the gross features of the oxygen arrangement, depicted as a hexagonal network in figure 4.3, but they still might be essential for a detailed comparison of the materials. As indicated by our representation in terms of O-V-O slabs we find alternating vanadium and oxygen layers along a_{prut} in both compounds. While the oxygen layers of the sesquioxide are almost flat, we are confronted with a distinct buckling in the case of the dioxide, which is caused by an elongated apical V-O distance, see the discussion in chapter 3. The buckling of the oxygen layers is associated with the fact that octahedral faces oriented perpendicular to a_{prut} in V_2O_3 adopt a tilt in the VO_2 case. Nonetheless, the characteristic feature distinguishing vanadium dioxide and sesquioxide certainly is their different pattern of filling the oxygen network with vanadium atoms.

Having developed a common description of the structures belonging to the end members of the vanadium Magnéli series it is now easy to expand this point of view to the remaining

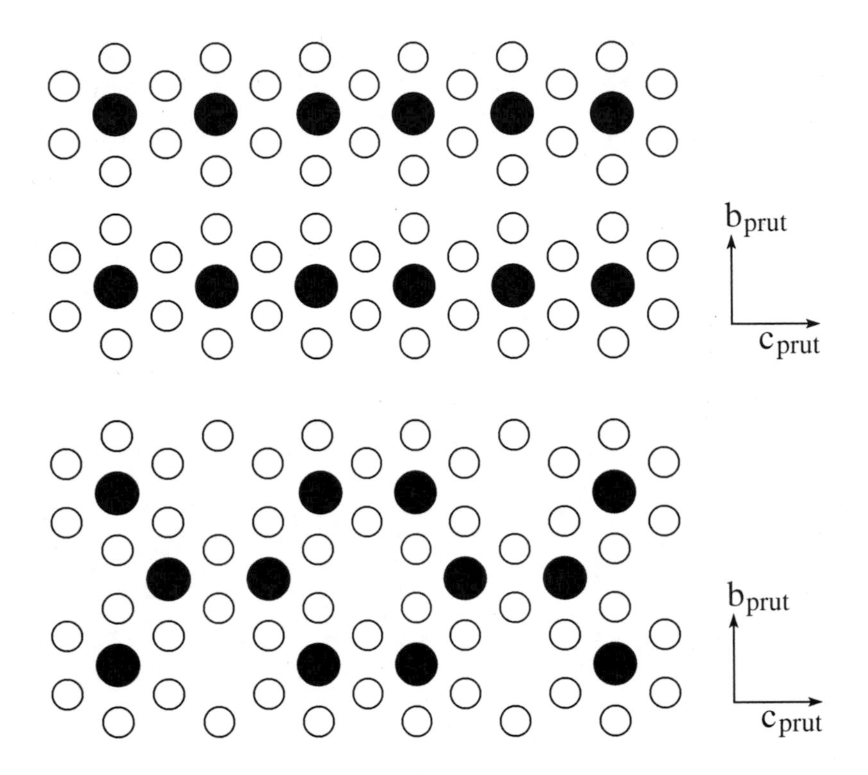

Figure 4.3: *Projection parallel to a_{prut} of a O-V-O sandwich-like slab cut out of the crystal structures of VO_2 (top) and V_2O_3 (bottom). Large and small spheres represent vanadium and oxygen atoms, respectively. Due to the projection the oxygen octahedra form a regular hexagonal network. For VO_2 infinite chains of vanadium atoms run along the pseudorutile c_{prut}-axis, where the oxygen octahedra are connected by common edges. The projection of an adjacent O-V-O slab results in a similar configuration – but empty and filled octahedra are exchanged. Thus the metal atoms in VO_2 lack nearest neighbours in the a_{prut}-direction. For V_2O_3 finite chains of two vanadium atoms run along the c_{prut}-axis. The projection of an adjacent O-V-O slab reveals a shift by one octahedral site along c_{prut}. As a consequence, the vanadium atoms exhibit exactly one nearest vanadium neighbour in the a_{prut}-direction.*

compounds. For that purpose the structures of V_4O_7 and V_6O_{11} are shown schematically in figure 4.4 – in full analogy to the representations of VO_2 and V_2O_3 in figure 4.3. Except for a slightly different buckling of the oxygen layers, which is illustrated for V_4O_7 in figure 4.5, the above mentioned regular three-dimensional oxygen network forms the basis of all Magnéli phases. Hence the hexagonal arrangement of the oxygen atoms in the projection of the O-V-O sandwich-like slabs does not change. Instead, the way of filling the oxygen

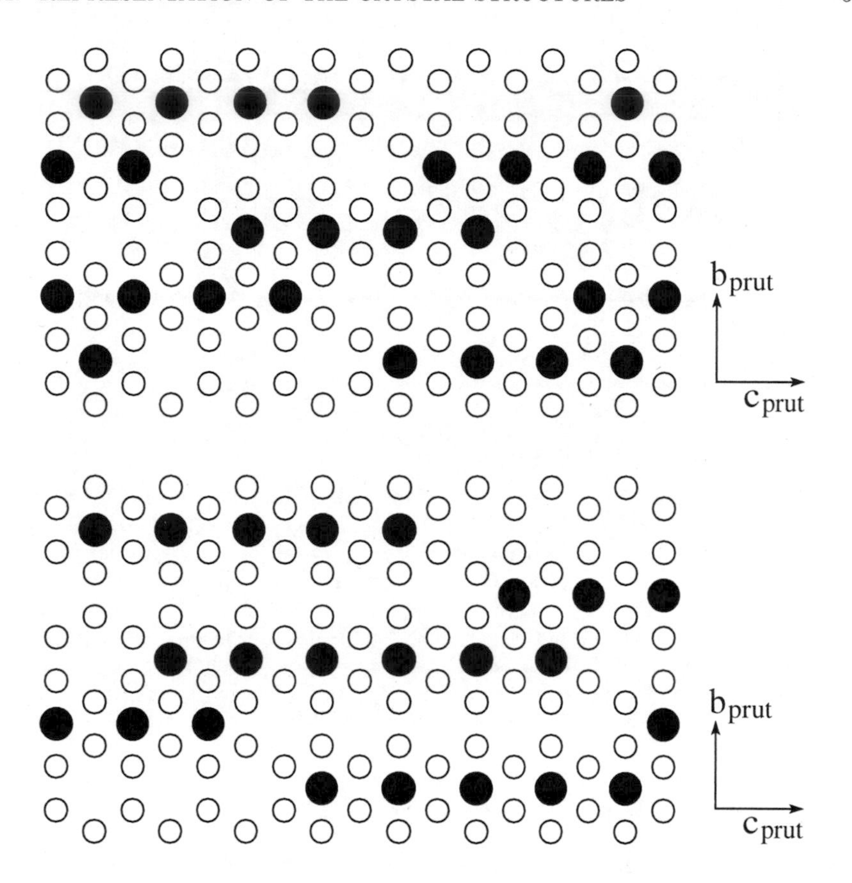

Figure 4.4: *Projection parallel to a_{prut} of a O-V-O sandwich-like slab cut out of the crystal structures of V_4O_7 (top) and V_6O_{11} (bottom). Large and small spheres denote vanadium and oxygen atoms, respectively. As in the cases of vanadium dioxide and sesquioxide the oxygen octahedra form a regular hexagonal network and the octahedral sites are partially occupied by metal atoms, giving rise to vanadium chains along c_{prut}. The chains consist of $n = 4/n = 6$ atoms, separated by $n - 1 = 3/n - 1 = 5$ empty octahedra. Along b_{prut} the chains overlap by roughly half the intrachain V-V distance. The projection of an adjacent O-V-O slab results in a similar configuration – but the vanadium sublattice is shifted along c_{prut} by $n - 1 = 3/n - 1 = 5$ octahedral sites. As a consequence, chain end atoms exhibit one nearest vanadium neighbour in the a_{prut}-direction, whereas center atoms reveal none.*

network with vanadium atoms characterizes the particular materials. As one immediately observes in figure 4.4 there are chains of vanadium atoms running along the c_{prut}-axis. In

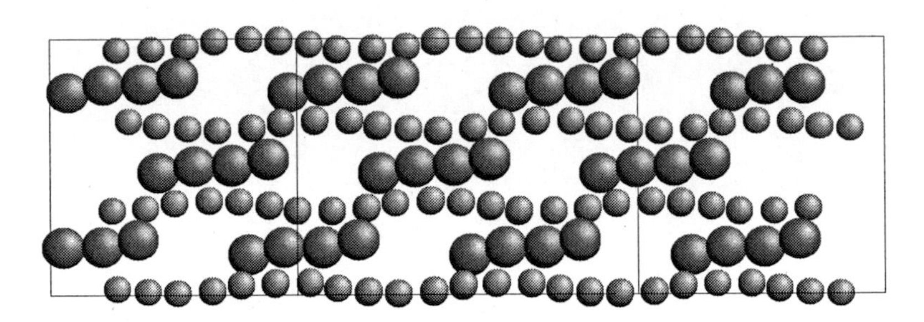

Figure 4.5: *Vanadium (large) and oxygen (small) layers in the case of the V_4O_7 structure – projection perpendicular to a_{prut}. A distinct buckling of the oxygen layers is clearly visible.*

the cases of V_4O_7 and V_6O_{11} they comprise four and six atoms, respectively. In general, we observe vanadium chains of length n in the Magnéli phase V_nO_{2n-1}. Looking back to the dioxide ($n = \infty$) and the sesquioxide ($n = 2$) this rule coincides with our finding of infinite chains and 2-chains, respectively. In a particular compound all metal chains have the same length – defined by the parameter n. Furthermore, exactly $n-1$ empty oxygen octahedra separate the chains in the c_{prut}-direction. Vanadium chains neighbouring along b_{prut} overlap by roughly half the in-chain V-V distance, thus giving rise to the characteristical chain end arrangement depicted in figure 4.4. In the whole Magnéli series this kind of end configuration is conserved and only the length of the chain center alters. For VO_2 there are actually no chain ends and the 2-chains of V_2O_3 contain no center atom.
The atomic arrangement represented by the projections of the O-V-O slabs in figure 4.4 stays unaltered for slabs neighbouring in the a_{prut}-direction. Only the vanadium sublattice shifts along the c_{prut}-axis by an amount of $n-1$ octahedral sites, i.e. $n-1$ hexagons in the sandwich-like projection. Thus the last vanadium atom of a metal chain is settled on top of the first atom of a chain in the adjacent slab. As a consequence, the chain end atoms exhibit exactly one nearest vanadium neighbour along the a_{prut}-axis, whereas each chain center atom is surrounded by two empty sites in this direction. All these considerations apply likewise to the other Magnéli phases. Chain center sites thus mirror the local atomic coordination known from VO_2 with nearest vanadium neighbours only in the chain direction. Instead, the coordination of the chain end sites resembles the crystal structure of V_2O_3 with a nearest vanadium neighbour both along a_{prut} and c_{prut}. Due to the varying length of the vanadium chains in a qualitatively unchanged oxygen sublattice the various Magnéli phases reveal different stoichiometric V:O ratios reaching from 1:2 in VO_2 to 2:3 in V_2O_3. Because the crystal structures of the different compounds are connected to each other simply by removing or inserting chain center atoms, the Magnéli series allows us to transfer a dioxide-type atomic arrangement step by step into a sesquioxide-type atomic arrangement. Using these systematics we can establish deeper insight into the interplay between VO_2 and V_2O_3-like regions of a crystal and their influence on the phase transition. Taking into account the mixed valent metal states of the Magnéli phases, the series transfers not only the crystal structure but also the electronic configuration of the dioxide

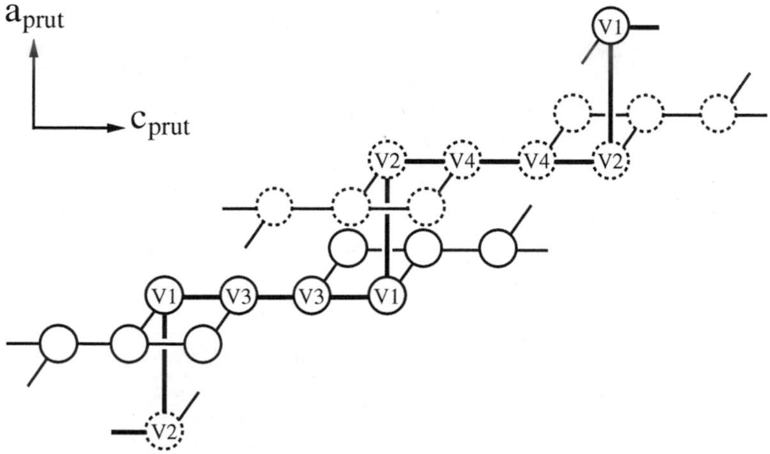

	V1–V3	V3–V3	V2–V4	V4–V4	V1–V2
T = 298 K	2.97	2.79	2.93	2.93	2.77
T = 120 K	3.03	2.67	2.83	3.03	2.78

Figure 4.6: *Crystal structure of V_4O_7. For simplicity this schematic view does not include the oxygen sublattice, which is a regular network of oxygen octahedra, qualitatively similar to the situation reported for VO_2 and V_2O_3. The vanadium atoms occupy octahedral sites and form chains along c_{prut} of length $n = 4$. Altogether there are four crystallographically inequivalent sites in the chains V1-V3-V3-V1 and V2-V4-V4-V2. Layers perpendicular to a_{prut} either comprise chains of the first or of the second kind. The table gives measured V-V distances (in Å) for both the metallic (298 K) and insulating (120 K) phase of V_4O_7 [95].*

(d^1) into that of the sesquioxide (d^2).

In the following sections two special Magnéli compounds are discussed in detail, namely V_4O_7 and V_6O_{11}. To define the notation of the different vanadium atoms figure 4.6 provides a schematic view of the V_4O_7 crystal structure where the regular network of oxygen octahedra has not been included for simplicity. One identifies vanadium 4-chains parallel to the c_{prut}-direction. Due to the mutual connections of the chains along a_{prut} a kind of stair-like vanadium arrangement arises. Stair endings resemble the corundum atomic coordination as displayed in figure 3.10, whereas intermediate plateaus are due to the rutile regions of the crystal. The longer the central part of the vanadium chains, the larger the separation of the stair endings and the dioxide-like character of the crystal. The illustration of V_6O_{11} in figure 4.7 is different from figure 4.6 due to the longer vanadium chains. We observe four crystallographically inequivalent vanadium sites in the case of V_4O_7 and six inequivalent sites for V_6O_{11}. In both structures these sites set up two different types of vanadium chains. For V_4O_7 the latter reveal the sequences V1-V3-V3-V1 and V2-V4-V4-

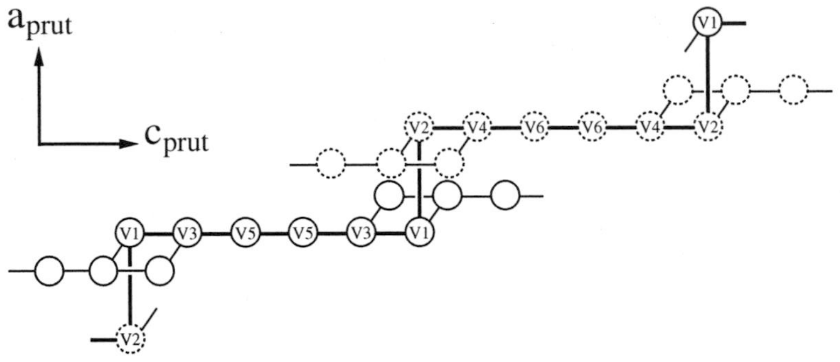

	V1–V3	V3–V5	V5–V5	V2–V4	V4–V6	V6–V6	V1–V2
T = 298 K	2.95	2.91	2.80	2.95	2.85	2.82	2.77
T = 20 K	2.90	2.99	2.79	3.06	2.62	3.30	2.77

Figure 4.7: *Crystal structure of* V_6O_{11}. *For simplicity this schematic view does not include the oxygen sublattice, which is a regular network of oxygen octahedra, qualitatively similar to the situation reported for* VO_2 *and* V_2O_3. *The vanadium atoms occupy octahedral sites and form chains along* c_{prut} *of length* $n = 6$. *Altogether there are six inequivalent sites in the chains V1-V3-V5-V5-V3-V1 as well as V2-V4-V6-V6-V4-V2. Layers perpendicular to* a_{prut} *either comprise chains of the first or of the second kind. The table gives measured V-V distances (in Å) for both the metallic (298 K) and insulating (20 K) phase of* V_6O_{11} *[96].*

V2, whereas the series V1-V3-V5-V5-V3-V1 and V2-V4-V6-V6-V4-V2 appear in V_6O_{11}. The inequivalent chains give rise to two different vanadium layers comprising either chains of the first or of the second kind, see figures 4.6 and 4.7. Within the vanadium sublattice these layers alternate along a_{prut} thus creating the shown arrangement.
The measured V-V distances given as insets in the above figures illustrate the dominating structural changes accompanying the transition of both V_4O_7 and V_6O_{11} [93–96]. At the phase transition of the former compound the dimerization in the 1-3 chains becomes stronger and an additional dimerization evolves in the 2-4 chains. The dimerization patterns are mutually reversed since the longer V-V distance appears at the chain ends and the chain center, respectively. The V1-V2 bond length is almost constant. For V_6O_{11} the changes in the 1-3-5 chains are rather small although a little dimerization is observed at low temperatures. In contrast, the 2-4-6 chains evolve a pronounced dimerization at the transition. Again the variation of the V1-V2 bond length is by far the smallest observed. A profound analysis of the structural changes induced by the MITs will be given in the subsequent sections where we relate LDA results to local structural properties. However, signatures of the VO_2-like dimerization are reflected by both V_4O_7 and V_6O_{11}.
As far as the question of choosing the unit cells of the Magnéli compounds is concerned, some differences are present in the literature – presumably due to the complicated atomic arrangement. A summary of different choices is given in the following. S. Andersson and

L. Jahnberg [83] defined primitive translations in terms of the rutile lattice, see equation (3.1). In doing so they distinguished the cases n odd

$$\begin{pmatrix} \mathbf{a}_A \\ \mathbf{b}_A \\ \mathbf{c}_A \end{pmatrix} = \begin{pmatrix} -1 & 0 & 1 \\ 1 & 1 & 1 \\ \frac{n-3}{2} & \frac{2-n}{2} & \frac{6-n}{2} \end{pmatrix} \begin{pmatrix} \mathbf{a}_R \\ \mathbf{b}_R \\ \mathbf{c}_R \end{pmatrix} \tag{4.2}$$

and n even

$$\begin{pmatrix} \mathbf{a}_A \\ \mathbf{b}_A \\ \mathbf{c}_A \end{pmatrix} = \begin{pmatrix} -1 & 0 & 1 \\ 1 & 1 & 1 \\ n-3 & 2-n & 6-n \end{pmatrix} \begin{pmatrix} \mathbf{a}_R \\ \mathbf{b}_R \\ \mathbf{c}_R \end{pmatrix} . \tag{4.3}$$

However, this proposal is not useful for comparison within the Magnéli series because for every parameter n the vanadium chains extend in different directions of the unit cell, and due to the two families of cell parameters. Y. Le Page and P. Strobel [97] thus proposed an alternative set of primitive translations with the c-axis parallel to the vanadium chains

$$\begin{pmatrix} \mathbf{a}_L \\ \mathbf{b}_L \\ \mathbf{c}_L \end{pmatrix} = \begin{pmatrix} -1 & 0 & 1 \\ 1 & 1 & 1 \\ 0 & 0 & 2n-1 \end{pmatrix} \begin{pmatrix} \mathbf{a}_R \\ \mathbf{b}_R \\ \mathbf{c}_R \end{pmatrix} . \tag{4.4}$$

While the unit cell of S. Andersson and L. Jahnberg is A-centered, the cell of Y. Le Page and P. Strobel is I-centered. In order to minimize the effort of electronic structure calculations both representations are not satisfactory as they are non-primitive. Consequently, we use the primitive unit cell introduced by H. Horiuchi *et al.* [98], which is given by the primitive translations

$$\begin{pmatrix} \mathbf{a}_H \\ \mathbf{b}_H \\ \mathbf{c}_H \end{pmatrix} = \begin{pmatrix} -1 & 0 & 1 \\ 1 & 1 & 1 \\ 0 & n-\frac{1}{2} & n-\frac{1}{2} \end{pmatrix} \begin{pmatrix} \mathbf{a}_R \\ \mathbf{b}_R \\ \mathbf{c}_R \end{pmatrix} . \tag{4.5}$$

Application of the above relation based on the parent rutile lattice is feasible whenever the crystal symmetry is triclinic with space group $P\bar{1}$ (C_i^1). This space group is characteristic for Magnéli phases with parameters $n = 4, ..., 9$ because the spacial arrangement of their vanadium chains in the oxygen network allows for only one lattice symmetry operation, the inversion [99]. The $n = 3$ compound V_3O_5 is an exception because it crystallizes in a body-centered monoclinic and a simple monoclinic lattice with space group $I2/c$ (C_{2h}^6) and $P2/c$ (C_{2h}^4) above and below the MIT, respectively [100, 101].

4.2 Structural and Calculational Details

The first vanadium Magnéli compound we discuss at length is $\mathbf{V_6O_{11}}$, see section 4.3. In doing so we can benefit from a suitable application of the knowledge we already acquired by our former considerations of VO_2. Therefore it is convenient to choose V_6O_{11} as the starting point of a comprehensive analysis of the whole Magnéli class. In contrast to the MIT of vanadium dioxide and the PM-AFI transition of vanadium sesquioxide, the MIT of V_6O_{11} at roughly 170 K is accompanied by a structural transformation preserving the crystal symmetry. Referring to x-ray studies by P. C. Canfield [96] the compound V_6O_{11} crystallizes in a triclinic lattice with space group $P\bar{1}$ (C_i^1). The vanadium as well as the oxygen atoms are located at the Wyckoff positions (2i): $\pm(x, y, z)$. For the triclinic lattice parameters the author reported the numbers $a_L = 5.449$ Å, $b_L = 7.010$ Å, $c_L = 31.437$ Å, $\alpha_L = 67.15°$, $\beta_L = 57.45°$, and $\gamma_L = 108.90°$ at a temperature of 298 K. Below the phase transition at 20 K he observed the parameters $a_L = 5.495$ Å, $b_L = 6.944$ Å, $c_L = 31.484$ Å, $\alpha_L = 67.40°$, $\beta_L = 57.13°$, and $\gamma_L = 108.61°$. All these values are related to the unit cell choice of Y. Le Page and P. Strobel as denoted in equation (4.4). We have to deal with primitive translations connected to the rutile lattice by ($n = 6$)

$$\begin{pmatrix} \mathbf{a}_L \\ \mathbf{b}_L \\ \mathbf{c}_L \end{pmatrix} = \begin{pmatrix} -1 & 0 & 1 \\ 1 & 1 & 1 \\ 0 & 0 & 11 \end{pmatrix} \begin{pmatrix} \mathbf{a}_R \\ \mathbf{b}_R \\ \mathbf{c}_R \end{pmatrix} . \tag{4.6}$$

The primitive translations of H. Horiuchi *et al.* take the form

$$\begin{pmatrix} \mathbf{a}_H \\ \mathbf{b}_H \\ \mathbf{c}_H \end{pmatrix} = \begin{pmatrix} -1 & 0 & 1 \\ 1 & 1 & 1 \\ 0 & \frac{11}{2} & \frac{11}{2} \end{pmatrix} \begin{pmatrix} \mathbf{a}_R \\ \mathbf{b}_R \\ \mathbf{c}_R \end{pmatrix} \tag{4.7}$$

and together with equation (4.6) we gain the relation

$$\begin{pmatrix} \mathbf{a}_H \\ \mathbf{b}_H \\ \mathbf{c}_H \end{pmatrix} = \begin{pmatrix} 1 & 0 & 0 \\ 0 & 1 & 0 \\ \frac{11}{2} & \frac{11}{2} & -\frac{1}{2} \end{pmatrix} \begin{pmatrix} \mathbf{a}_L \\ \mathbf{b}_L \\ \mathbf{c}_L \end{pmatrix} . \tag{4.8}$$

Coordinates given in the L-system can directly be transformed into the H-system

$$\begin{pmatrix} x \\ y \\ z \end{pmatrix}_H = \begin{pmatrix} 1 & 0 & 11 \\ 0 & 1 & 11 \\ 0 & 0 & -2 \end{pmatrix} \begin{pmatrix} x \\ y \\ z \end{pmatrix}_L + \begin{pmatrix} 1/2 \\ 0 \\ 1/2 \end{pmatrix}_L , \tag{4.9}$$

where the additional translation vector must be added because of different choices of the origin in both coordinate systems. With the help of equation (4.9) the positional parameters reported by P. C. Canfield for both the high and the low temperature configuration of V_6O_{11} yield the structural input for the band structure calculation, which is summarized in table 4.8. In addition to the six crystallographically inequivalent vanadium atoms there are eleven oxygen sites. As a consequence of the inversion symmetry of the crystal lattice the unit cell finally comprises 34 atoms. Further structural data for the metallic phase of V_6O_{11} were obtained by H. Horiuchi *et al.* [98]. However, due to their higher experimental accuracy the subsequent LDA calculations refer to the data of P. C. Canfield.
To allow for an adequate interpretation of the electronic structure results, it is important to preserve close relation between the representation of the V_6O_{11} structure and the parent rutile structure. We therefore arrange the primitive translations of the L-system in a Cartesian coordinate system so that the alignment of the oxygen octahedra resembles the

	High temperature structure			Low temperature structure		
Atom	x	y	z	x	y	z
V1	0.9931	0.4467	0.0624	0.9986	0.4633	0.0584
V2	0.9731	0.9264	0.0672	0.0081	0.9568	0.0612
V3	0.9652	0.4583	0.2422	0.9578	0.4379	0.2506
V4	0.9560	0.9463	0.2432	0.9362	0.9362	0.2484
V5	0.9765	0.4689	0.4170	0.9885	0.5086	0.4074
V6	0.9823	0.9777	0.4152	0.0582	0.0473	0.4036
O1	0.6888	0.4114	0.0332	0.6584	0.3744	0.0402
O2	0.3202	0.5570	0.0666	0.3173	0.5650	0.0658
O3	0.6845	0.4670	0.1090	0.6809	0.4685	0.1098
O4	0.3736	0.6180	0.1460	0.3464	0.5937	0.1508
O5	0.6930	0.4060	0.2134	0.6954	0.4057	0.2150
O6	0.3138	0.5883	0.2468	0.3360	0.6176	0.2396
O7	0.6697	0.4485	0.2968	0.6683	0.4469	0.2974
O8	0.3171	0.5339	0.3474	0.3161	0.5435	0.3468
O9	0.7039	0.4223	0.3898	0.6826	0.3998	0.3940
O10	0.3180	0.5976	0.4254	0.3510	0.6150	0.4224
O11	0.6864	0.4689	0.4718	0.6983	0.4710	0.4704

Table 4.8: *Atomic positions for the high and low temperature structure of V_6O_{11} as used in the band structure calculations. These data have been determined by P. C. Canfield [96] and the coordinates refer to the primitive translations proposed by H. Horiuchi et al. [98].*

rutile lattice. For keeping similarity to equations (4.6) to (4.9) we presume the primitive translations in Cartesian coordinates to fulfill

$$\mathbf{a}_L = \begin{pmatrix} -a_1 \\ 0 \\ a_3 \end{pmatrix}, \quad \mathbf{b}_L = \begin{pmatrix} a_1 \\ b_2 \\ a_3 \end{pmatrix}, \quad \mathbf{c}_L = \begin{pmatrix} c_1 \\ c_2 \\ c_3 \end{pmatrix}. \tag{4.10}$$

An application of the elementary relations

$$a_L^2 = a_1^2 + a_3^2, \quad b_L^2 = a_1^2 + b_2^2 + a_3^2, \quad \text{and} \quad \mathbf{a}_L \cdot \mathbf{b}_L = a_L b_L \cos\gamma_L \tag{4.11}$$

immediately yields

$$a_{1/3}^2 = \frac{1}{2}\left(a_L^2 \mp a_L b_L \cos\gamma_L\right) \quad \text{and} \quad b_2^2 = a_L^2 - a_L^2. \tag{4.12}$$

Furthemore, from the expressions

$$\mathbf{a}_L \cdot \mathbf{c}_L = a_L c_L \cos\beta_L \quad \text{and} \quad \mathbf{b}_L \cdot \mathbf{c}_L = b_L c_L \cos\alpha_L \tag{4.13}$$

we obtain

$$c_{1/3} = \frac{b_L c_L \cos\alpha_L \mp a_L c_L \cos\beta_L - b_2 c_2}{2a_{1/3}}, \tag{4.14}$$

which implies a quadratic equation for c_2 because $c_L^2 = c_1^2 + c_2^2 + c_3^2$. Solving this equation completes the calculation of the unit cell in the L-system and transforming the calculated primitive translations via equation (4.8) yields the unit cell in the H-system. For the high temperature structure of V_6O_{11} at 298 K we find ($A = 8.6665\, a_B$)

$$\mathbf{a}_H = A \begin{pmatrix} -1.0000 \\ 0.0000 \\ 0.6417 \end{pmatrix}, \quad \mathbf{b}_H = A \begin{pmatrix} 1.0000 \\ 0.9616 \\ 0.6417 \end{pmatrix}, \quad \mathbf{c}_H = A \begin{pmatrix} -0.0053 \\ 5.4629 \\ 3.6352 \end{pmatrix}. \tag{4.15}$$

The low temperature structure at 20 K results in ($A = 8.6981\, a_B$)

$$\mathbf{a}_H = A \begin{pmatrix} -1.0000 \\ 0.0000 \\ 0.6521 \end{pmatrix}, \quad \mathbf{b}_H = A \begin{pmatrix} 1.0000 \\ 0.9224 \\ 0.6521 \end{pmatrix}, \quad \mathbf{c}_H = A \begin{pmatrix} -0.0074 \\ 5.3417 \\ 3.7636 \end{pmatrix}. \tag{4.16}$$

Here $a_B = 0.529177$ Å is the Bohr radius and the primitive translations refer to Cartesian coordinates. The corresponding positional parameters of the vanadium and oxygen atoms are denoted in table 4.8.

All the following LDA calculations have been performed using the scalar relativistic augmented spherical wave (ASW) method, as introduced at the end of the second chapter. In particular, a parametrization of the exchange correlation potential given by S. H. Vosko, L. Wilk, and M. Nusair [20] was applied. The ASW scheme is based on the atomic sphere approximation and hence models the crystal potential by means of spherically symmetric atomic potentials. It is required that the atomic spheres fill the space of the unit cell. For open crystal structures problems can arise because space filling only due to atom centered spheres may lead to large overlap. As a consequence, so-called empty spheres, i.e. pseudo atoms without nuclei, have to be introduced. They are used to correctly model the shape of the crystal potential in large voids. The collection of both physical and empty spheres leads to an artificial close-packed structure. However, since the potential of the whole set of spheres ought to represent the crystal potential as exactly as possible it is a challenge to find optimal empty sphere positions and optimal radii for the real and empty spheres. For this purpose the sphere geometry optimization algorithm described in [102] is most helpful and efficient.

By inserting altogether 35 empty spheres from 18 crystallographically inequivalent classes into the triclinic unit cell of high temperature V_6O_{11} it is possible to keep the linear overlap of real spheres below 18%. Simultaneously, the overlap of any pair of real and empty spheres is smaller than 23%. In the case of the low temperature structure 21 inequivalent classes with altogether 40 empty spheres allow for reducing the overlaps to less than 18% and 23%, respectively. Summing up, the unit cell entering the LDA calculation contains 69 spheres for the high and 74 spheres for the low temperature configuration. The radii of the physical vanadium and oxygen spheres are summarized in table 4.9, which additionally denotes the valence charges arising from the LDA band structure calculation. These results will be discussed in detail in the following section. For both V_6O_{11} structures the basis sets taken into account in the secular matrix (2.72) comprise V $4s$, $4p$, $3d$ and O $2s$, $2p$ orbitals. In addition, V $4f$ and O $3d$ states are included as tails of the aforementioned orbitals (for details on the ASW method see [28]). In order to complete these basis sets

	High temperature structure		Low temperature structure	
Atom	Radius	Charge	Radius	Charge
V1	2.1866	2.5090	2.2303	2.6179
V2	2.1587	2.7155	2.2278	2.8276
V3	2.2833	2.5240	2.2158	2.5534
V4	2.2755	2.7564	2.2248	2.7939
V5	2.3440	2.7775	2.2248	2.5213
V6	2.3437	2.7354	2.1856	2.6201
O1	1.9684	4.1741	1.9423	4.2072
O2	1.7883	4.0164	1.8240	3.9466
O3	1.9680	4.3682	1.9699	4.4187
O4	2.0093	4.3624	1.9420	4.2381
O5	1.7654	3.8853	1.8219	3.8832
O6	1.8752	4.0164	1.8201	3.9683
O7	1.8673	4.0673	1.7943	3.9711
O8	1.8908	4.0278	1.8394	4.0294
O9	1.8585	4.0833	1.8568	4.0498
O10	1.8828	4.1106	1.8121	3.9479
O11	1.9170	4.0959	1.8166	3.8850

Table 4.9: *Atomic sphere radii of vanadium and oxygen (in a_B) as well as calculated LDA valence charges (V 3d or O 2p) for both the high and the low temperature phase of V_6O_{11}.*

we add empty sphere states, which we determine with respect to the spacial extension of the spheres and the total charge occupying them. The configurations reach from $1s$, $(2p)$ to $1s$, $2p$, $3d$, $4f$, $(5g)$, where states in parentheses again enter as tails of other orbitals. During the course of the LDA calculation the Brillouin zone is sampled using increasing numbers of **k**-points in the irreducible wedge. In this manner one ensures convergence of the results with respect to the fineness of the **k**-space grid. For both the high and the low temperature calculation the number of **k**-points in the irreducible wedge of the Brillouin zone was initially 108 and increased to 256, 864, and 2048. As convergence criteria for the self-consistency of the charge density, the deviations of the atomic charges and the total energy of subsequent iterations were required to be less than 10^{-8} electrons and 10^{-8} Ryd, respectively (1 Ryd $\approx$ 13.6 eV). These requirements are a standard for the calculations of the following sections.

After having considered V_6O_{11} we turn to $\mathbf{V_4O_7}$, which is a compound characterized by its comparatively small vanadium chains of length $n = 4$. It is therefore suitable for investigating the electronic features of the sesquioxide-like chain end atoms and their influence on the MIT. Having at hand some fundamental ideas developed in the context of V_6O_{11} we will be able to analyze the MIT of V_4O_7. Afterwards we will establish important implications for the frequently discussed MITs of V_2O_3, see section 4.4. In the same manner as V_6O_{11} the MIT of V_4O_7 at 250 K is accompanied by structural distortions, which keep the space group $P\bar{1}$ (C_i^1).

	High temperature structure			Low temperature structure		
Atom	x	y	z	x	y	z
V1	0.9806	0.9383	0.1008	0.9773	0.9299	0.1043
V2	0.9946	0.4509	0.0964	0.9781	0.4382	0.1008
V3	0.9662	0.9613	0.3742	0.9508	0.9512	0.3757
V4	0.9914	0.4905	0.3658	0.0146	0.5021	0.3608
O1	0.6927	0.4137	0.0514	0.6932	0.4151	0.0512
O2	0.3237	0.5551	0.1036	0.3191	0.5383	0.1112
O3	0.6825	0.4643	0.1716	0.6808	0.4658	0.1718
O4	0.3763	0.6232	0.2254	0.3804	0.6214	0.2254
O5	0.6893	0.4112	0.3328	0.7018	0.4240	0.3278
O6	0.3044	0.5683	0.3910	0.2793	0.5257	0.4068
O7	0.6605	0.4326	0.4700	0.6616	0.4218	0.4728

Table 4.10: *Atomic positions for the high and low temperature structure of V_4O_7 as used in the band structure calculations. These data trace back to J.-L. Hodeau and M. Marezio [95]; the coordinates refer to the primitive translations proposed by H. Horiuchi et al. [98].*

Structural refinements by J.-L. Hodeau and M. Marezio [95] revealed both the vanadium and the oxygen atoms in V_4O_7 at the Wyckoff positions (2i): $\pm(x, y, z)$. Above the phase transition (298 K) the authors found the lattice parameters $a_A = 5.509$ Å, $b_A = 7.008$ Å, $c_A = 12.256$ Å, $\alpha_A = 95.10°$, $\beta_A = 95.17°$, and $\gamma_A = 109.25°$. The corresponding values for the low temperature phase (120 K) amount to $a_A = 5.503$ Å, $b_A = 6.997$ Å, $c_A = 12.256$ Å, $\alpha_A = 94.86°$, $\beta_A = 95.17°$, and $\gamma_A = 109.39°$. The numbers refer to the unit cell choice of S. Andersson and L. Jahnberg, see equation (4.3). Hence the primitive translations are connected to the rutile lattice by ($n = 4$)

$$\begin{pmatrix} \mathbf{a}_A \\ \mathbf{b}_A \\ \mathbf{c}_A \end{pmatrix} = \begin{pmatrix} -1 & 0 & 1 \\ 1 & 1 & 1 \\ 1 & -2 & 2 \end{pmatrix} \begin{pmatrix} \mathbf{a}_R \\ \mathbf{b}_R \\ \mathbf{c}_R \end{pmatrix} . \tag{4.17}$$

With equation (4.5) the primitive translations of H. Horiuchi *et al.* are given by

$$\begin{pmatrix} \mathbf{a}_H \\ \mathbf{b}_H \\ \mathbf{c}_H \end{pmatrix} = \begin{pmatrix} -1 & 0 & 1 \\ 1 & 1 & 1 \\ 0 & \frac{7}{2} & \frac{7}{2} \end{pmatrix} \begin{pmatrix} \mathbf{a}_R \\ \mathbf{b}_R \\ \mathbf{c}_R \end{pmatrix} . \tag{4.18}$$

Combining this relation with equation (4.17) yields

$$\begin{pmatrix} \mathbf{a}_H \\ \mathbf{b}_H \\ \mathbf{c}_H \end{pmatrix} = \begin{pmatrix} 1 & 0 & 0 \\ 0 & 1 & 0 \\ 2 & \frac{5}{2} & -\frac{1}{2} \end{pmatrix} \begin{pmatrix} \mathbf{a}_A \\ \mathbf{b}_A \\ \mathbf{c}_A \end{pmatrix} \tag{4.19}$$

and we are able to transform coordinates from the A-system into the H-system

$$\begin{pmatrix} x \\ y \\ z \end{pmatrix}_H = \begin{pmatrix} 1 & 0 & 4 \\ 0 & 1 & 5 \\ 0 & 0 & -2 \end{pmatrix} \begin{pmatrix} x \\ y \\ z \end{pmatrix}_A + \begin{pmatrix} 1/2 \\ 1/2 \\ 1/2 \end{pmatrix}_A . \tag{4.20}$$

	High temperature structure		Low temperature structure	
Atom	Radius	Charge	Radius	Charge
V1	2.2002	2.7937	2.1388	2.7457
V2	2.2349	2.6572	2.3468	2.7529
V3	2.3224	2.7329	2.1999	2.7430
V4	2.3433	2.7245	2.3436	2.6778
O1	1.9931	4.3397	2.0096	4.3259
O2	1.8278	4.0266	1.9267	4.1770
O3	1.9883	4.4309	1.9759	4.4115
O4	2.0465	4.3530	2.0226	4.3440
O5	1.7994	3.9182	1.7492	3.8328
O6	1.8993	4.0359	1.8357	3.9669
O7	1.9164	4.1097	1.9045	4.0616

Table 4.11: *Atomic sphere radii of both vanadium and oxygen (in a_B) and calculated LDA valence charges (V 3d or O 2p) for the high as well as the low temperature phase of V_4O_7.*

The additional translation vector has been added in order to account for different choices of the origin in the coordinate systems. Using equation (4.20) the positional parameters obtained by J.-L. Hodeau and M. Marezio for high and low temperature V_4O_7 result in the values summarized in table 4.10, which enter the LDA band structure calculations. Less accurate investigations of the crystal structure trace back to H. Horiuchi *et al.* [93] and M. Marezio *et al.* [94]. Altogether, one obtains four crystallographically inequivalent vanadium sites and seven additional oxygen sites. Due to the inversion symmetry of the lattice the V_4O_7 unit cell contains 22 atoms.
Starting with the lattice parameters determined by J.-L. Hodeau and M. Marezio we set up the triclinic unit cell of V_4O_7 in Cartesian coordinates. Because of similar definitions of the A-system and the L-system, see equations (4.3) and (4.4), we can proceed in full analogy to the case of V_6O_{11}. Intead of the lattice constants a_L, b_L, c_L, α_L, β_L and γ_L the corresponding values a_A, b_A, c_A, α_A, β_A and γ_A have to enter equations (4.10) to (4.14). Hence we end with a unit cell in the A-system and via equation (4.19) with a unit cell in the H-system. For the high temperature structure at 298 K this yields ($A = 8.7702\, a_B$)

$$\mathbf{a}_H = A \begin{pmatrix} -1.0000 \\ 0.0000 \\ 0.6396 \end{pmatrix}, \quad \mathbf{b}_H = A \begin{pmatrix} 1.0000 \\ 0.9333 \\ 0.6396 \end{pmatrix}, \quad \mathbf{c}_H = A \begin{pmatrix} 0.0071 \\ 3.4280 \\ 2.3282 \end{pmatrix}. \tag{4.21}$$

Moreover, the low temperature configuration at 120 K results in ($A = 8.7691\, a_B$)

$$\mathbf{a}_H = A \begin{pmatrix} -1.0000 \\ 0.0000 \\ 0.6374 \end{pmatrix}, \quad \mathbf{b}_H = A \begin{pmatrix} 1.0000 \\ 0.9313 \\ 0.6374 \end{pmatrix}, \quad \mathbf{c}_H = A \begin{pmatrix} 0.0054 \\ 3.4199 \\ 2.3141 \end{pmatrix}. \tag{4.22}$$

These primitive translations refer to Cartesian coordinates; they are used with positional parameters from table 4.10.

Similar to the procedure in the case of V_6O_{11} we insert additional augmentation spheres into the structure of V_4O_7 to account for its openness and to properly model the crystal potential. For both the high and the low temperature phase it suffices to use 13 crystallographically inequivalent classes and place altogether 24 empty spheres in the triclinic unit cell. This allows us to keep the linear overlap of real spheres below 19% and the overlap of pairs of real and empty spheres below 24%. Unit cells entering the LDA band structure calculation finally comprise 46 spheres. Table 4.11 denotes the radii of the physical vanadium and oxygen spheres. For later use calculated valence charges are also included. The basis sets taken into account in the secular matrix comprise the same orbitals as for V_6O_{11}. Moreover, the technical details of the LDA calculations for high and low temperature V_4O_7 are identical to those used for the latter material.

By comparing $\mathbf{V_7O_{13}}$ to $\mathbf{V_8O_{15}}$ the influence of the chain center sites on the electronic properties of the Magnéli phases will be analyzed quantitatively in section 4.5. This comparison is of special interest since the former compound is the only member of the Magnéli series, which does not undergo an MIT. In contrast, V_8O_{15} exhibits an MIT at about 70 K accompanied by structural distortions. V_7O_{13} as well as both V_8O_{15} phases crystallize in the well known triclinic space group $P\bar{1}$ (C_i^1).
Referring to investigations of P. C. Canfield [96] one observes the vanadium and oxygen atoms in V_7O_{13} and V_8O_{15} at the Wyckoff positions (2i): $\pm(x, y, z)$. The following lattice parameters reported by this author are related to the unit cell choice of Y. Le Page and P. Strobel as established in equation (4.4). In the case of V_7O_{13} one finds $a_L = 5.439$ Å, $b_L = 7.013$ Å, $c_L = 37.161$ Å, $\alpha_L = 67.04°$, $\beta_L = 57.46°$, and $\gamma_L = 108.92°$. For the room temperature structure of V_8O_{15} these constants amount to $a_L = 5.431$ Å, $b_L = 7.017$ Å, $c_L = 42.896$ Å, $\alpha_L = 66.84°$, $\beta_L = 57.55°$, and $\gamma_L = 108.94°$, while the low temperature configuration is characterized by the values $a_L = 10.892$ Å, $b_L = 6.980$ Å, $c_L = 85.907$ Å, $\alpha_L = 66.83°$, $\beta_L = 57.38°$, and $\gamma_L = 108.67°$. The latter unit cell is exceptional because it accounts for a superstructure evolving in V_8O_{15} at low temperatures, which complicates the considerations. However, for the high temperature structures we may proceed in full analogy to the previous discussion of V_6O_{11} and thus give only the results. Atomic coordinates have to be transformed from the L into the H-system. For this purpose we use in the case of V_7O_{13} the relation

$$\begin{pmatrix} x \\ y \\ z \end{pmatrix}_H = \begin{pmatrix} 1 & 0 & 13 \\ 0 & 1 & 13 \\ 0 & 0 & -2 \end{pmatrix} \begin{pmatrix} x \\ y \\ z \end{pmatrix}_L \tag{4.23}$$

and for high temperature V_8O_{15}

$$\begin{pmatrix} x \\ y \\ z \end{pmatrix}_H = \begin{pmatrix} 1 & 0 & 15 \\ 0 & 1 & 15 \\ 0 & 0 & -2 \end{pmatrix} \begin{pmatrix} x \\ y \\ z \end{pmatrix}_L + \begin{pmatrix} 1/2 \\ 0 \\ 1/2 \end{pmatrix}_L . \tag{4.24}$$

Applying these transformations to the positional parameters measured by P. C. Canfield leads to the structural input for the LDA calculation, which is summarized in table 4.12. Alternative, though less accurate, crystallographic studies of V_7O_{13} at room temperature were carried out by H. Horiuchi *et al.* [98]. Finally, we denote the primitive translations

	Metallic V_7O_{13}			Metallic V_8O_{15}		
Atom	x	y	z	x	y	z
V1	0.9969	0.9507	0.5522	0.9990	0.4517	0.0452
V2	0.9708	0.4237	0.5576	0.9665	0.9187	0.0508
V3	0.9604	0.9526	0.7064	0.9636	0.4573	0.1786
V4	0.9510	0.4376	0.7081	0.9475	0.9312	0.1818
V5	0.9634	0.9532	0.8565	0.9524	0.4378	0.3118
V6	0.9796	0.4756	0.8520	0.9723	0.9666	0.3070
V7	0.0000	0.0000	0.0000	0.9782	0.4782	0.4372
V8	0.0000	0.5000	0.0000	0.0009	0.0011	0.4334
O1	0.3042	0.0901	0.0154	0.6901	0.4148	0.0238
O2	0.6901	0.9770	0.0506	0.3201	0.5583	0.0490
O3	0.3115	0.0273	0.1011	0.6851	0.4671	0.0802
O4	0.6912	0.9078	0.1379	0.3691	0.6124	0.1082
O5	0.2989	0.0840	0.1687	0.6934	0.4055	0.1568
O6	0.6827	0.9664	0.2053	0.3132	0.5891	0.1816
O7	0.3290	0.0493	0.2490	0.6719	0.4530	0.2176
O8	0.6873	0.9117	0.2906	0.3120	0.5250	0.2572
O9	0.3074	0.0944	0.3191	0.6994	0.4127	0.2880
O10	0.6304	0.8868	0.3754	0.3074	0.5941	0.3140
O11	0.3158	0.0354	0.4072	0.6901	0.4778	0.3452
O12	0.6815	0.9432	0.4433	0.3096	0.5211	0.3904
O13	0.3101	0.0854	0.4724	0.6930	0.4036	0.4212
O14				0.2981	0.5882	0.4480
O15				0.6940	0.4866	0.4762

Table 4.12: *Atomic positions for the metallic structures of both V_7O_{13} and V_8O_{15} as used in the band structure calculations. These data have been determined by P. C. Canfield [96] and the coordinates refer to the primitive translations proposed by H. Horiuchi et al. [98].*

(in Cartesian coordinates) belonging to the structural data in table 4.12. For V_7O_{13} they are given by ($A = 8.6547\ a_B$)

$$\mathbf{a}_H = A\begin{pmatrix} -1.0000 \\ 0.0000 \\ 0.6406 \end{pmatrix}, \quad \mathbf{b}_H = A\begin{pmatrix} 1.0000 \\ 0.9666 \\ 0.6406 \end{pmatrix}, \quad \mathbf{c}_H = A\begin{pmatrix} -0.0046 \\ 6.4668 \\ 4.2749 \end{pmatrix} \tag{4.25}$$

and for high temperature V_8O_{15} we obtain ($A = 8.6459\ a_B$)

$$\mathbf{a}_H = A\begin{pmatrix} -1.0000 \\ 0.0000 \\ 0.6396 \end{pmatrix}, \quad \mathbf{b}_H = A\begin{pmatrix} 1.0000 \\ 0.9712 \\ 0.6396 \end{pmatrix}, \quad \mathbf{c}_H = A\begin{pmatrix} -0.0102 \\ 7.4675 \\ 4.9097 \end{pmatrix}. \tag{4.26}$$

In the case of the insulating V_8O_{15} modification P. C. Canfield used a non-primitive unit

	Insulating V_8O_{15}					
Atom	x	y	z	$\tilde{x}$	$\tilde{y}$	$\tilde{z}$
V1	0.9288	0.2657	0.0432	0.4319	0.2694	0.0428
V2	0.4424	0.7839	0.0548	0.9235	0.7645	0.0448
V3	0.9449	0.2124	0.1780	0.4442	0.2155	0.1824
V4	0.4576	0.7140	0.1820	0.9711	0.7226	0.1868
V5	0.9881	0.1208	0.3196	0.4724	0.1260	0.3012
V6	0.4525	0.6129	0.3040	0.9427	0.6006	0.3008
V7	0.9831	0.0488	0.4416	0.4908	0.0439	0.4436
V8	0.4720	0.5272	0.4268	0.0134	0.5613	0.4392
O1	0.6877	0.0797	0.0224	0.1944	0.0849	0.0260
O2	0.6006	0.3958	0.0464	0.1080	0.4078	0.0528
O3	0.4160	0.5297	0.0812	0.9079	0.5234	0.0784
O4	0.3691	0.8779	0.1088	0.8616	0.8705	0.1076
O5	0.7216	0.0052	0.1572	0.2300	0.0075	0.1584
O6	0.1286	0.3578	0.1804	0.6320	0.3614	0.1816
O7	0.4584	0.4774	0.2184	0.9511	0.4801	0.2160
O8	0.4407	0.8366	0.2616	0.9320	0.8191	0.2552
O9	0.7587	0.9324	0.2904	0.2514	0.9311	0.2872
O10	0.1543	0.2885	0.3124	0.6771	0.3070	0.3192
O11	0.4758	0.3903	0.3456	0.9765	0.3908	0.3440
O12	0.4637	0.7592	0.3904	0.9703	0.7653	0.3920
O13	0.7779	0.8476	0.4204	0.2742	0.8447	0.4184
O14	0.1926	0.2244	0.4480	0.6924	0.2284	0.4484
O15	0.5003	0.3161	0.4756	0.0014	0.3195	0.4760

Table 4.13: *Atomic positions for the insulating structure of V_8O_{15} as applied in the LDA band structure calculations. The data trace back to P. C. Canfield [96] and the coordinates refer to the L''-system of primitive coordinates as defined in the text. Every metallic site gives rise to two inequivalent sites in the insulating case, which are distinguished by tildes.*

cell given by the primitive translations

$$\begin{pmatrix} \mathbf{a}_{L'} \\ \mathbf{b}_{L'} \\ \mathbf{c}_{L'} \end{pmatrix} = \begin{pmatrix} -2 & 0 & 2 \\ 1 & 1 & 1 \\ 0 & 0 & 30 \end{pmatrix} \begin{pmatrix} \mathbf{a}_R \\ \mathbf{b}_R \\ \mathbf{c}_R \end{pmatrix} . \tag{4.27}$$

They are related to the unit cell choice of Y. Le Page and P. Strobel, as defined in equation (4.4), by a doubling of both $\mathbf{a}_L$ and $\mathbf{c}_L$. To set up this cell in Cartesian coordinates we therefore have to modify the initial assumption (4.10) to

$$\mathbf{a}_{L'} = \begin{pmatrix} -2a_1 \\ 0 \\ 2a_3 \end{pmatrix} , \quad \mathbf{b}_{L'} = \begin{pmatrix} a_1 \\ b_2 \\ a_3 \end{pmatrix} , \quad \mathbf{c}_{L'} = \begin{pmatrix} c_1 \\ c_2 \\ c_3 \end{pmatrix} . \tag{4.28}$$

	Metallic V_7O_{13}		Metallic V_8O_{15}	
Atom	Radius	Charge	Radius	Charge
V1	2.1752	2.6856	2.1780	2.6323
V2	2.1470	2.5531	2.1455	2.7006
V3	2.2785	2.6930	2.2918	2.6544
V4	2.2521	2.5878	2.2476	2.6812
V5	2.2885	2.6140	2.2689	2.7533
V6	2.3325	2.7843	2.3382	2.6668
V7	2.3486	2.6903	2.3579	2.7017
V8	2.3657	2.7684	2.3623	2.6763

	Insulating V_8O_{15}			
Atom	Radius	Charge	Radius	Charge
V1	2.1387	2.5244	2.1204	2.4989
V2	2.1176	2.6540	2.3199	2.8019
V3	2.3203	2.5383	2.1768	2.7219
V4	2.2764	2.6839	2.1241	2.5339
V5	2.1482	2.5851	2.3145	2.5971
V6	2.3600	2.7499	2.3197	2.6284
V7	2.2323	2.7182	2.2980	2.7176
V8	2.3763	2.7557	2.2158	2.5337

Table 4.14: *Atomic sphere radii of vanadium (in a_B) and calculated LDA valence charges (V 3d) for metallic V_7O_{13}/V_8O_{15} as well as insulating V_8O_{15}. The left side of the second table refers to the coordinates (x, y, z) in the superstructure of V_8O_{15}, the right to $(\tilde{x}, \tilde{y}, \tilde{z})$.*

Afterwards we proceed in analogy to equations (4.11) to (4.14). However, the outcome of this calculation has to be transformed into a primitive unit cell using the relations

$$\mathbf{a}_{L''} = \mathbf{a}_{L'}\,, \quad \mathbf{b}_{L''} = \mathbf{b}_{L'}\,, \quad \mathbf{c}_{L''} = -\frac{1}{4}\mathbf{a}_{L'} + \frac{1}{2}\mathbf{b}_{L'} - \frac{1}{4}\mathbf{c}_{L'}\,. \tag{4.29}$$

A further transformation into the H-system is neither reasonable nor possible. Hence we found the primitive translations entering the LDA calculation for the V_8O_{15} low temperature phase ($A = 8.6420\ a_B$)

$$\mathbf{a}_{L''} = A \begin{pmatrix} -2.0000 \\ 0.0000 \\ 1.2933 \end{pmatrix}, \quad \mathbf{b}_{L''} = A \begin{pmatrix} 1.0000 \\ 0.9547 \\ 1.2933 \end{pmatrix}, \quad \mathbf{c}_{L''} = A \begin{pmatrix} 0.9820 \\ 0.7188 \\ -4.6900 \end{pmatrix}. \tag{4.30}$$

In the next step the positional parameters reported by P. C. Canfield for low temperature V_8O_{15} have to be transformed into the L''-system. We use equation (4.29) to establish

$$\begin{pmatrix} x \\ y \\ z \end{pmatrix}_{L''} = \begin{pmatrix} 1 & 0 & -1 \\ 0 & 1 & 2 \\ 0 & 0 & -4 \end{pmatrix} \begin{pmatrix} x \\ y \\ z \end{pmatrix}_{L'}. \tag{4.31}$$

Applying the above transformation leads to the calculational input summarized in table 4.13. Since each vanadium site of the metallic unit cell gives rise to two inequivalent sites in the insulating superstructure we introduce the notation V1,...V8,$\tilde{\mathrm{V1}}$,...,$\tilde{\mathrm{V8}}$ to account for this interrelation.
Bearing in mind the inversion symmetry inherent in the crystal structure, the unit cell of V_7O_{13} comprises 40 atoms, which belong to seven vanadium and thirteen oxygen classes. Furthermore, the unit cell of high temperature V_8O_{15} contains 46 atoms in eight vanadium and fifteen oxygen classes. Due to the superstructure of low temperature V_8O_{15} we have sixteen inequivalent vanadium and thirty oxygen sites. The unit cell comprises altogether 92 atoms. To obtain an artificial close-packed structure we again complement the physical sites with auxiliary empty spheres. For V_7O_{13} it suffices to dispose 41 empty spheres from 21 inequivalent classes to keep the linear overlap of real spheres below 18%. For high/low temperature V_8O_{15} we apply 50/92 spheres from 26/52 classes to ensure a linear overlap of real spheres smaller than 19%/20%. Furthermore, the overlap of any pair of real and empty spheres exceeds the maximum of the real sphere overlap by less than 5 percentage points. Summing up, the unit cell entering the LDA calculation comprises 81 spheres for V_7O_{13}. Moreover, we must take into account 96/184 spheres in the case of V_8O_{15}. Sphere radii and calculated valence charges of the vanadium sites are summarized in table 4.14. The technical details of the LDA calculation are the same as those used for the previous Magnéli phases. In particular, we take into account the standard basis sets in the secular matrix. To reduce the computing time, the self-consistency convergence criterium for the total energy was set to 10^{-6} Ryd in the case of low temperature V_8O_{15}.

4.3 Dimerization and Localization in V_6O_{11}

In this section we study the MIT of V_6O_{11} by means of electronic structure calculations based on the density functional theory and the local density approximation, applying the augmented spherical wave method. Changes of the electronic structure at the transition are discussed in relation to the structural transformations occuring simultaneously. The study will benefit from our unified representation of the crystal structures of the Magnéli phases as well as of VO_2 and V_2O_3. We will succeed in grouping the electronic bands of V_6O_{11} into states behaving similarly to the dioxide or the sesquioxide. Hence it is possible to analyze the phase transitions of the different compounds on the basis of a common frame of reference, which helps us to gain insight into the delicate interplay of electron-lattice coupling and electronic correlations and their influence on the MITs. Parts of the subsequent findings have been discussed in a previous paper [103].
In analogy to the investigation of vanadium dioxide and sesquioxide we start our considerations with a survey of the electronic states arising from the LDA calculations for high and low temperature V_6O_{11}. The findings are displayed in figure 4.15 in the same energy interval used in figures 3.5 and 3.11 for VO_2 and V_2O_3, respectively. The electronic bands in both cases are depicted along selected high symmetry lines in the first Brillouin zone of the triclinic lattice, which is shown in figure 4.16 in order to define the high symmetry points. The low temperature antiferromagnetism of V_6O_{11} is not considered for reasons discussed at the beginning of this chapter.

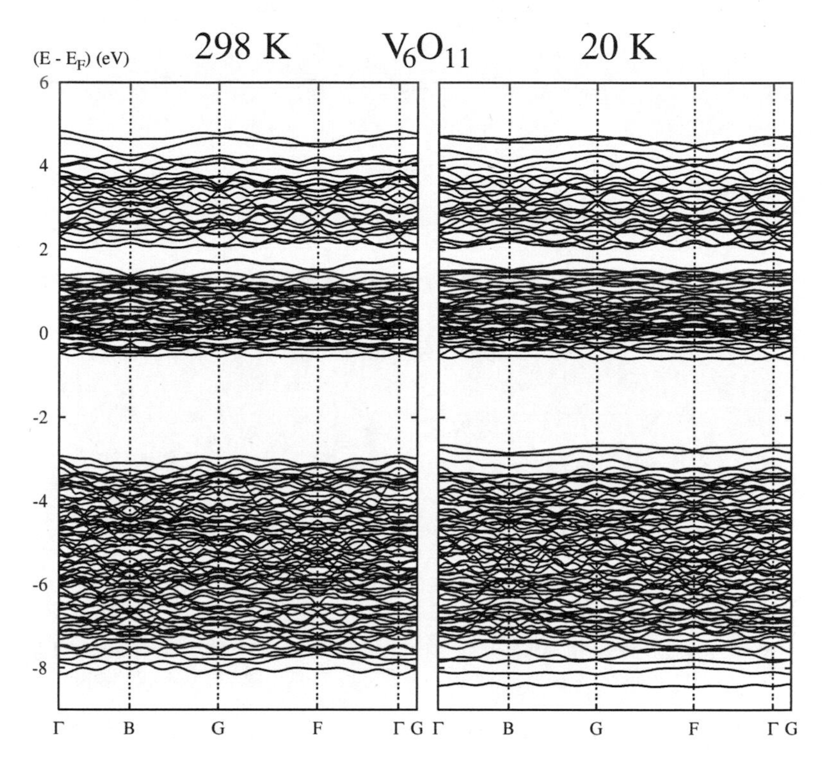

Figure 4.15: *Electronic bands of high and low temperature V_6O_{11} displayed along selected symmetry lines in the first Brillouin zone of the triclinic lattice as depicted in figure 4.16.*

Having a closer look at figure 4.15 we find the electronic bands reveal dispersion throughout the first Brillouin zone. Since the unit cell of V_6O_{11} comprises quite a few atoms we are confronted with a large number of electronic states, giving rise to a multitude of bands in small energy ranges. Moreover, as is obvious from figure 4.15, the electronic states are affected by strong hybridization effects. An analysis of orbitally weighted electronic bands consequently is not reasonable since it provides no insight into the effects of the particular atomic states on the overall electronic structure. In comparison to the discussion of vanadium dioxide and sesquioxid we have therefore lost a very important tool for investigating the relationship between structural features and electronic properties. Consequently, only a very sophisticated analysis of the local DOS at the individual atomic sites can pave the way for understanding the transition of V_6O_{11}. Similar to VO_2 and V_2O_3 we come across three groups of bands in figure 4.15. In the high temperature case they are settled in the energy ranges from -8.2 eV to -2.9 eV, from -0.7 eV to 1.8 eV, and from 2.0 eV to 4.8 eV. On entering the low temperature phase the lowest group broadens considerably, reaching from -8.4 eV to -2.7 eV, whereas both other groups stay mainly unaltered. Due to sub-

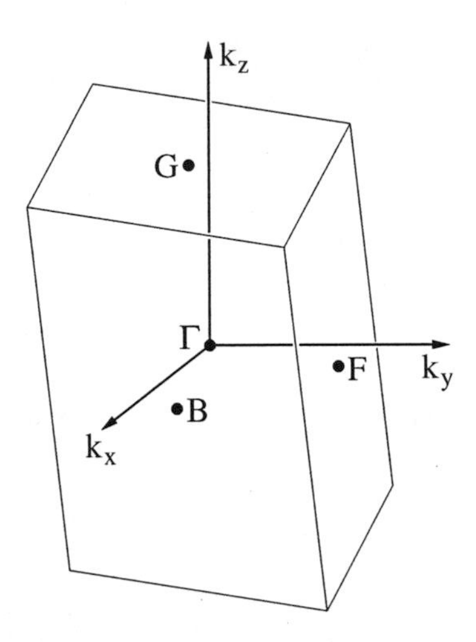

Figure 4.16: *First Brillouin zone of the triclinic lattice. Labels mark high symmetry points.*

stantial structural similarities to both the dioxide and the sesquioxide it is not surprising that the electronic structure of V_6O_{11} resembles the expectations of the molecular orbital picture as dicussed for the former compounds. The characteristic σ and π-type overlap of the V $3d$ and O $2p$ states again gives rise to bonding-antibonding split molecular orbitals satisfying the energetical order σ-π-π^*-σ^*. Here the bonding states are identified with the lowest group of electronic bands, the π^* states with the middle, and the σ^* states with the highest group. Since each vanadium site is surrounded by an oxygen octahedron, the σ^* and π^* bands reveal e_g^σ and t_{2g} symmetry, respectively. The number of bands covered by the three groups is easily calculated as there are 12 vanadium and 22 oxygen atoms in the unit cell of both the high and the low temperature structure. Hence we have $12 \times 2 = 24$ vanadium $3d$ e_g^σ, $12 \times 3 = 36$ vanadium $3d$ t_{2g}, and $22 \times 3 = 66$ oxygen $2p$ bands, see figure 4.15.

Figure 4.17 complements the electronic band structure with partial V $3d$ and O $2p$ densities of states calculated for high and low temperature V_6O_{11}. The three groups observed in the band structure reappear as three distinct contributions in the DOS. In agreement with the molecular orbital picture and the findings for VO_2 and V_2O_3 the energetically lowest structure primarily traces back to O $2p$ orbitals, whereas the other groups mainly originate from V $3d$ states. Remarkable contributions of vanadium and oxygen in energy ranges dominated by the respective other states are due to hybridization effects between V $3d$ and O $2p$ states. In regions corresponding to σ-type V-O overlap they consequently are somewhat stronger. This is true for the e_g^σ energy range and for the lower half of the O $2p$ dominated region. In comparison to the results for vanadium dioxide and sesquioxide there is very good accordance of the gross features of the DOS. Thus we can transfer our knowledge from chapter 3 to the electronic structure of V_6O_{11}. As summarized in table 4.1

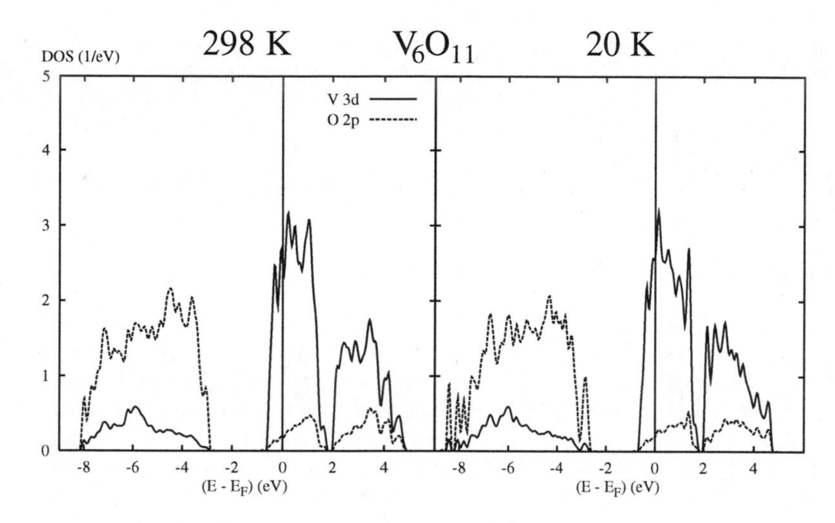

Figure 4.17: *Partial V 3d and O 2p densities of states (DOS) per vanadium atom resulting from the high (298 K) and low (20 K) temperature crystal structure of* V_6O_{11} *are compared.*

the formal $3d$ charge of the vanadium sites in V_6O_{11} is 1.33 electrons, which establishes an intermediate position between the single electron of VO_2 and the two electrons of V_2O_3. The characteristic differences in the electron count yield a different filling level of the V $3d$ t_{2g} orbitals. Figures 3.7, 3.13, and 4.17 confirm the conclusion by different energetical positions of the t_{2g} group of states in relation to the Fermi energy. Comparing the LDA results arising from the high and the low temperature V_6O_{11} structure reveals hardly any difference. No energy gap is observed, but the structural modifications accompanying the MIT leave the DOS at the Fermy level mainly unaffected. This is not surprising because for neither the insulating phase of VO_2 nor the AFI phase of V_2O_3 did the LDA calculation succeed in reproducing an energy gap. Again it is reasonable to attribute this failure to the shortcomings of the local density approximation. However, as we have learned from the dioxide, the shortcomings do not prevent us from understandig the mechanism of the MIT. Especially since we aim at investigating the relation between the electronic states and the local atomic environments as well as the modification of this relationship at the MIT, the LDA limitations are not important for the following considerations.
In the next step we decompose the V $3d$ t_{2g} group of states into its symmetry components. Therefore figures 4.18 to 4.20 give site-projected partial $d_{x^2-y^2}$, d_{yz}, and d_{xz} densities of states for all six crystallographically inequivalent vanadium sites V1,...,V6. For each site the presentation of the data refers to the local rotated reference frame, which was defined in the discussion of the rutile structure. The rutile reference frame is useful since the local octahedral coordination of the metal sites in V_6O_{11} resembles the rutile arrangement, see the preceding sections. The gross features of all the site-projected densities of states are similar to those reported for VO_2 and V_2O_3. In particular, one obtains two groups of an-

tibonding states resulting from V-O overlap combined with crystal field splitting. First, there is the energy range from -0.7 eV to 1.8 eV where the t_{2g} bands dominate. Second, we observe t_{2g} contributions also between 2.0 eV and 4.8 eV, i.e. in the e_g^σ region. Larger admixtures in the case of V_6O_{11}, compared to VO_2, point to an increased deviation from an ideal octahedral coordination of the vanadium sites. The octahedral distortions agree more with those known from the sesquioxide. Figure 4.17 shows additional contributions of oxygen states in the energy interval of figures 4.18 to 4.20. In the vicinity of the Fermi energy they amount to less than 10% of the total DOS but are still indicative of covalent bonding between vanadium and oxygen atoms.
The major part of the structural transformation of V_6O_{11} at 170 K is concerned with the strong V4-V6 dimerization in the 2-4-6 vanadium chain, see figure 4.7. While the V4-V6 bond length decreases from 2.85 Å to 2.62 Å, the V2-V4 and V6-V6 distances grow from 2.95 Å to 3.06 Å and from 2.82 Å to 3.30 Å, respectively. As a consequence, two isolated V4-V6 vanadium pairs arise and the chain end atoms (V2) separate from the rest of the vanadium chain. According to figure 4.7 dimerization effects lead to a shortened V5-V5 bond length in the 1-3-5 chain both above and below the MIT. Obviously, in the insulating phase the V5-V5 pair separates even more from the evolving V1-V3 pairs. All these modifications of the crystal structure resemble the pairing effects known from the Peierls distortion in the monoclinic low temperature phase of VO_2. Due to the finite vanadium chains in the case of V_6O_{11} it is not surprising that we find more complicated distortion patterns. Various influences of the sesquioxide-like chain ends affect metal-metal pairing in the whole chain. The different behaviour of the inequivalent vanadium chains will be discussed in a later section.
The monoclinic structure of VO_2 is characterized not only by metal-metal pairing but also by zigzag-type in-plane displacements of the vanadium atoms parallel to the local z-axis, thus parallel to the diagonals of the rutile basal planes. While the vanadium sites in the rutile structure coincide with the centers of the surrounding oxygen octahedra, they are shifted away from these positions by 0.20 Å in the distorted configuration – hence giving rise to an antiferroelectric mode. Equivalent shifts away from the centers of their oxygen octahedra are reported for the sites V3, V4, V5, and V6 in V_6O_{11}. The distance between vanadium atoms and octahedral centers grows from 0.16 Å to 0.29 Å (V3), from 0.17 Å to 0.23 Å (V4), from 0.07 Å to 0.23 Å (V5), and from 0.04 Å to 0.35 Å (V6), respectively. Actually, the major part of the shifts is oriented perpendicular to the c_{prut}-axis. Although the vanadium sites in the high temperature phase do not coincide with the octhedral centers, the relative changes at the MIT are well known from the dioxide. Interestingly, the high temperature lateral displacement is larger for the intermediate sites V3 and V4 than for the center atoms V5 and V6. This is due to a slight rotation of the vanadium chains away from the c_{prut}-axis, which is found for each Magnéli phase. Because of the rotation the chain end atoms V1 and V2 shift away from their nearest neighbours along a_{prut}. The latter distortion mirrors the vanadium anti-dimerization observed in V_2O_3 parallel to the c_{hex}-axis. Here the displacement of the vanadium sites with respect to the centers of the oxygen octahedra is 0.18 Å in the PM and 0.21 Å in the AFI phase. In the case of V_6O_{11} the displacements are increased to 0.32 Å for the metallic and 0.35 Å for the insulating configuration. Thus the antiferroelectric-like shifts of the atoms V1 and V2 are present in both phases with minor changes. Even with respect to the details of the local distortion,

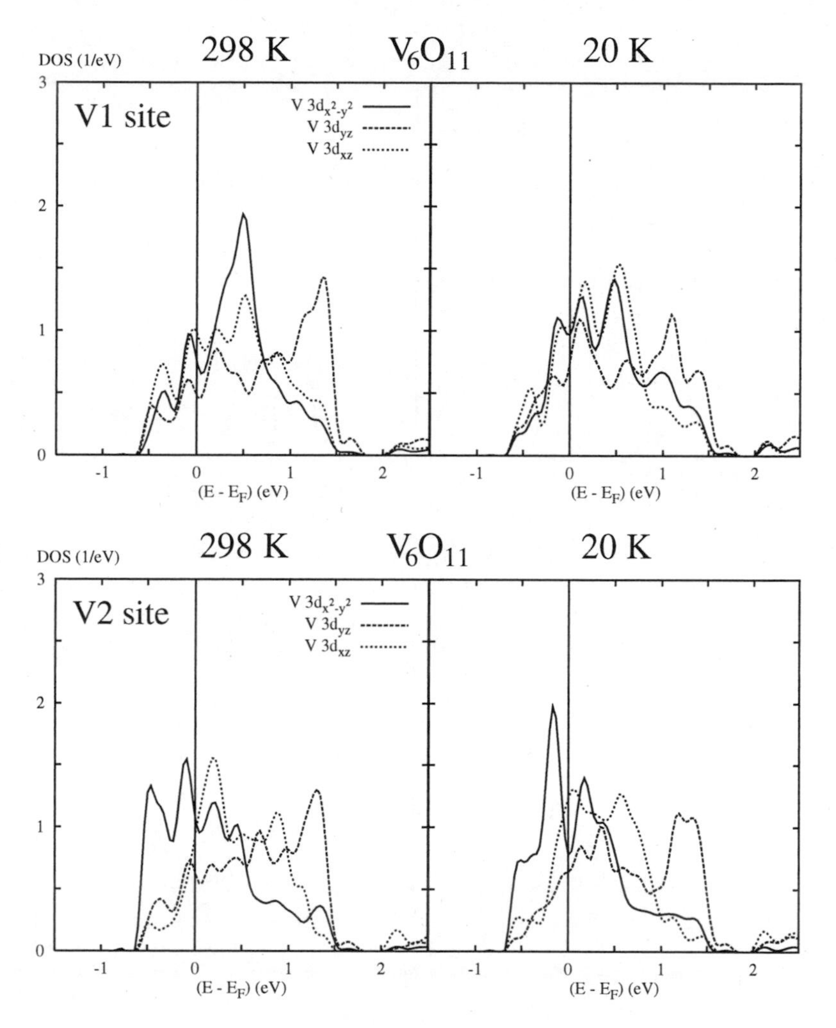

Figure 4.18: *Site-projected partial V 3d t_{2g} densities of states (DOS) per metal atom for the high and the low temperature crystal structure of V_6O_{11}: sites V1 and V2. The crystal structure is depicted in figure 4.7 and the orbitals refer to the local rotated reference frame.*

chain centers and ends behave as the parent structures VO_2 and V_2O_3, respectively. Going into more detail, we turn to the t_{2g} states of the vanadium atoms V4, V5, and V6, which are involved in strong dioxide-like displacements. Thus the site-projected densities of states are similar to one another and resemble the DOS of rutile and monoclinic VO_2,

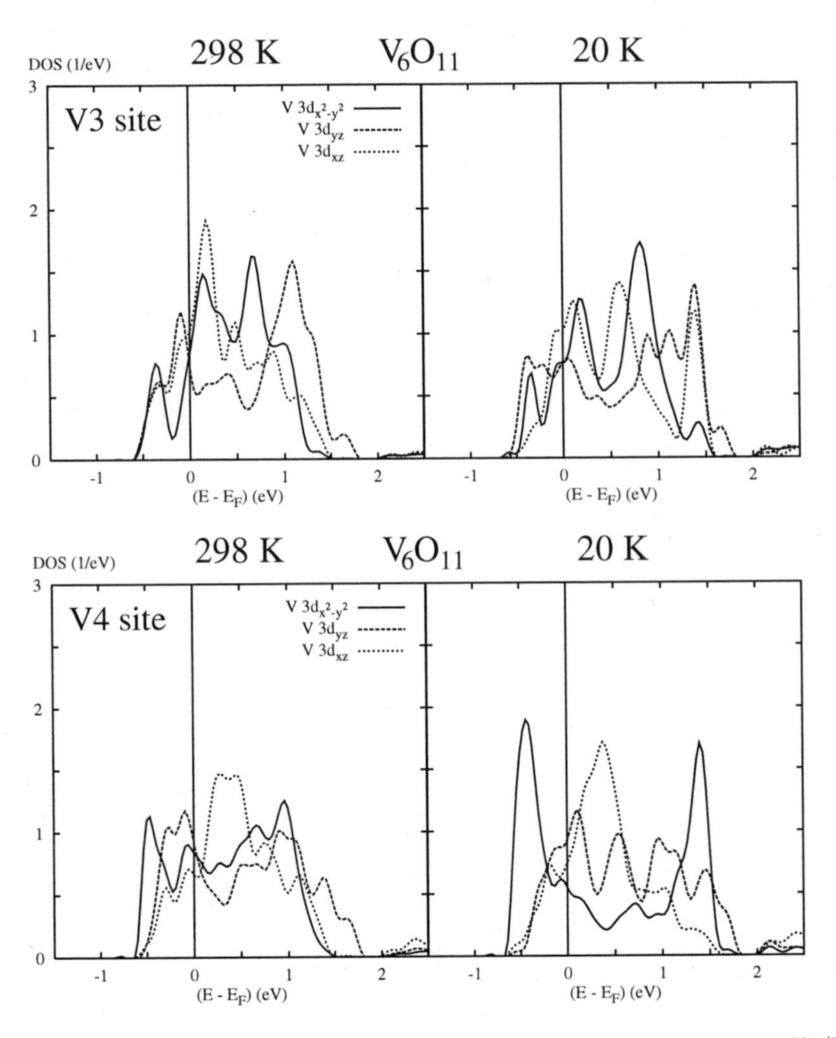

Figure 4.19: *Same representation as used in figure 4.18, but for the vanadium sites V3/V4.*

compare figure 3.7. Note the weakly indicated two peak structure of the high temperature V4 $d_{x^2-y^2}$ DOS due to σ-type metal-metal bonding along the vanadium chains. In the low temperature phase the V4-V6 dimerization causes an increased splitting of the DOS into bonding and antibonding branches located at energies of roughly $-0.5\,\mathrm{eV}$ and $1.4\,\mathrm{eV}$. At the same time the d_{yz} and d_{xz} densities of states undergo energetical upshifts due to the antiferroelectric displacement of the V4 atoms perpendicular to c_{prut}. As outlined before,

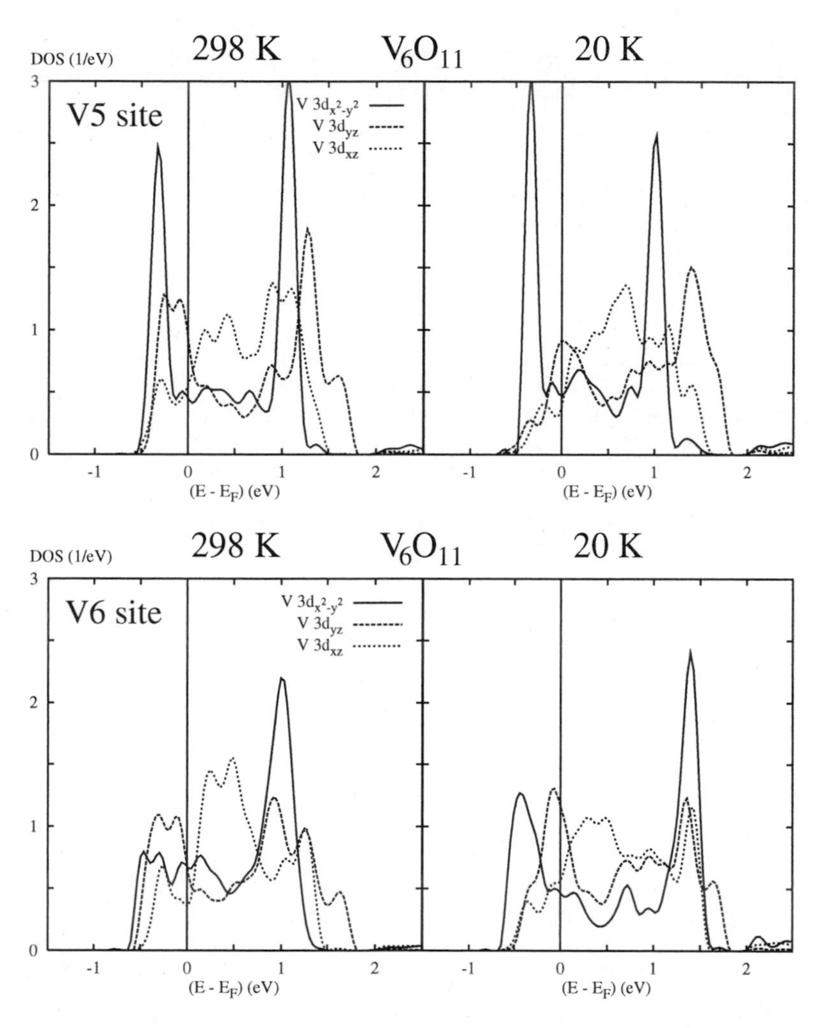

Figure 4.20: *Same representation as used in figure 4.18, but for the vanadium sites V5/V6.*

such distortions raise the overlap between the V 3d and O 2p orbitals and hence the π-π^* splitting. Equivalent to the situation reported for the dioxide, the energetical separation between the $d_{x^2-y^2} = d_{\parallel}$ band and the remaining e_g^π states is increased but not complete in the low temperature phase.

The situation is similar for the V6 atoms as they participate in the V4-V6 pairs. Thus the bonding and antibonding branches of the $d_{x^2-y^2}$ DOS appear at the same energies as for

the V4 site. However, in the low temperature phase the splitting is more pronounced due to additional V6-V6 bonding. As in the case of V4 the energetical upshift of the d_{yz} and d_{xz} densities of states is easily observed, especially directly below the Fermi energy. While the d_{xz} DOS of V4/V6 gives rise to a broad single peak, we obtain a distinct splitting of the d_{yz} DOS in contributions centered at about -0.2 eV and 1.0 eV for the high temperature structure. The splitting is understood by taking into account metal-metal bonding parallel to b_{prut}. Across octahedral faces typical vanadium sequences V2-V6-V4 appear in this direction, which becomes obvious from figure 4.4. Because of the remarkable peak in the low temperature V6 d_{yz} DOS at roughly 1.4 eV, not present in the respective V4 DOS but reappearing in the V2 DOS, we may assume an increased V2-V6 bonding parallel to b_{prut} at low temperatures. The $d_{x^2-y^2}$ and d_{yz} densities of states of atom V5 show strong bonding-antibonding splitting, which hardly changes at the phase transition. The shape of the $d_{x^2-y^2}$ DOS is easily understandable in terms of the constant V5-V5 bonding. In contrast, the splitting of the d_{yz} DOS traces back to metal-metal overlap along b_{prut}. The energetical upshift of these states at the MIT is pronounced and has the same origin as discussed for V4 and V6. In total, the partial V4, V5, and V6 densities of states display characteristic features known from the dimerization and the zigzag-type displacements in monoclinic VO_2 and thus can be understood in terms of the latter compound.
In contrast to the center sites V4, V5, and V6 the chain end atoms (V1,V2) are characterized by a sesquioxide-like local environment. They are particularly involved in metal-metal bonding across shared octahedral faces connecting the 1-3-5 to the 2-4-6 layers, but they are not subject to dimerization along c_{prut}. As a consequence, except for small peaks and shoulders near -0.5 eV and from 1.0 eV to 1.3 eV (reminiscent of the chain center $d_{x^2-y^2}$ densities of states) the V1/V2 $d_{x^2-y^2}$ DOS consists of a single broad peak extending from -0.2 eV to 0.7 eV. The subpeaks observed in this energy range miss counterparts in the DOS of any neighbouring vanadium site. Since such peaks are indicative of metal-metal bonding we conclude, consistent with the reported V-V bond lengths, that the chain end atoms are well separated from the remainder of the chains. Hence the $d_{x^2-y^2}$ states may be regarded as localized. While chain center $d_{x^2-y^2}$ orbitals predominantly mediate V-V overlap along c_{prut}, this interaction is suppressed for V1 and V2 due to geometrical reasons. Hence chain end $d_{x^2-y^2}$ orbitals behave as d_{xz} states, which generally give rise to a reduced (π-type) metal-metal overlap parallel to the chains. In both phases the center of weight of the V1 $d_{x^2-y^2}$ DOS is noticeably higher than that of the V2 $d_{x^2-y^2}$ DOS. Due to vanadium next neighbours along a_{prut} the d_{yz} orbitals are important for interpreting chain end atoms because these states may cause V-V bonding parallel to both a_{prut} and b_{prut}. Actually, the V1 and V2 d_{yz} states show bonding-antibonding splitting above and below the MIT. For the insulating modification the separation is more pronounced with peaks directly above the Fermi level and around 1.2 eV. However, so far we are not able to distinguish between the influences of the two possible metal-metal interactions. Whether the shape of the d_{yz} DOS indicates bonding along a_{prut} or b_{prut} or both will be dealt with later. However, this question is of interest since the a_{prut}-direction of V_6O_{11} corresponds to the c_{hex}-axis of the corundum structure. Hence the splitting of the d_{yz} states is equivalent to the a_{1g} band splitting in V_2O_3. From the analysis of the V_6O_{11} chain end atoms we probably can gain better insight into the MIT of the sesquioxide. The partial V1 and V2 t_{2g} densities of states qualitatively resemble each other due to similar sesquioxide-like

surroundings and can obviously be interpreted in accordance with V_2O_3.
Atom V3 is exceptional since its t_{2g} DOS can neither be interpreted by means of dioxide-like dimerization in the vanadium chains nor by sesquioxide-like metal-metal overlap along a_{prut}. The second fact is simply due to the lacking of adjacent vanadium sites, which allow for overlap between the 1-3-5 and 2-4-6 layers. Nevertheless, all three components of the t_{2g} partial DOS are similar to those of the atoms V1 and V2. The $d_{x^2-y^2}$ and d_{xz} densities of states consist of broad single peaks. Reminiscent of bonding and antibonding branches in the corresponding V5 DOS we find satellite peaks at -0.4 eV and a shoulder at 1.0 eV in the high temperature phase. The compact shape of the d_{xz} DOS again is not surprising due to the small V-V overlap this orbital is involved in, not only for V3 but also for the other vanadium atoms. In contrast, the d_{yz} partial DOS displays a distinct double peak structure in both phases, indicative of bonding-antibonding splitting. At first glance this finding is difficult to understand as geometrically no vanadium partner along a_{prut} exists, which could participate in the V-V bonding. However, this puzzling situation is resolved by reconsidering the possibility of V-V interaction perpendicular to the chains and simultaneously perpendicular to the a_{1g}-direction, namely along b_{prut}. As stated before, we are confronted with finite vanadium sequences consisting of three atoms (V1-V5-V3), where the distance between neighbouring sites corresponds to the longer rutile lattice constant. To be more precise, the bond lengths amount to 4.58 Å/4.54 Å (V1-V5) and 4.45 Å/4.40 Å (V5-V3) for the high/low temperature configuration. Furthermore, the associated values in the V2-V6-V4 sequences are 4.54 Å/4.57 Å (V2-V6) and 4.49 Å/4.38 Å (V6-V4). Because each vanadium site in V_6O_{11} participates in one of the above sequences, the σ-type overlap parallel to b_{prut} and the resulting bonding-antibonding splitting of the d_{yz} states is present not only for atom V3. We found almost the same effect in VO_2, see figure 3.7. The above findings have important implications for the interpretation of the V1/V2 d_{yz} DOS. In particular, the interactions parallel to a_{prut} and b_{prut} add to the bonding-antibonding splitting of these states. The respective antibonding peaks of V1 and V2 are merged in the high temperature phase but split up at the MIT. Due to the V-V overlap along b_{prut} atom V3 resembles the electronic properties of the chain end sites, although its coordination does not allow for a_{1g}-like bonding. Hence V-V bonding across the vanadium layers is not the only source for a splitting of the d_{yz} DOS. Despite the differing bond lengths hardly any difference between σ-type metal-metal overlap via octahedral faces within and perpendicular to the vanadium layers is found. In accordance with results of I. S. Elfimov *et al.* [76] this indicates the importance of V-V couplings other than the a_{prut}-like for the shape of the a_{1g} bands in V_2O_3.
While the overlap of vanadium and oxygen places the V $3d$ t_{2g} states near the Fermi level, the detailed electronic features of the orbitals, and hence the MIT of V_6O_{11}, are fundamentally influenced by the local metal-metal coordination. Small variations of the oxygen sublattice at the phase transition barely affect the shape of any t_{2g} DOS, whereas modifications of the V-V bond lengths influence the electronic properties. The latter are typical for VO_2 or V_2O_3 depending on the local environment of the particular vanadium atom. Therefore it is reasonable to regard the partial densities of states as local quantities. Near the chain centers (atoms V4, V5, and V6) the dimerization and the antiferroelectric-like displacements via strong electron-lattice interaction cause a splitting of the $d_{x^2-y^2}$ DOS and an energetical upshift of the e_g^π states at the MIT. For atom V5 the former effect is

already present at high temperatures. Together, the chain center sites behave in analogy to VO_2. In contrast, near the chain ends (atoms V1, V2, and V3) the $d_{x^2-y^2}$ orbitals are characterized by strongly reduced metal-metal overlap and thus reveal a localized nature. The shape of the partial $d_{x^2-y^2}$ DOS is closely related to the d_{xz} DOS. Due to the strong localization electronic correlations may play a major role for the chain end $d_{x^2-y^2}$ states. Being subject only to V-V overlap within a vanadium layer the partial V3 d_{yz} DOS shows bonding-antibonding splitting similar to the atoms V1 and V2, which have an additional nearest neighbour across the layers. Therefore the in-plane V-V interaction is at least as important as the perpendicular overlap. In conclusion, the sesquioxide-like regions of the V_6O_{11} crystal appear to be susceptible to electronic correlations for geometrical reasons. Structural variations at the phase transition leave all three components of the t_{2g} partial DOS mainly unchanged, thus reflecting close relations to the sesquioxide. Consequently, the MIT of V_6O_{11} is interpreted as resulting from the combination of electron-lattice interaction and electronic correlations. Due to different atomic arrangements in VO_2 and V_2O_3-like regions neither an embedded Peierls instability, as known from the dioxide, nor sesquioxide-like correlations can account for the MIT of V_6O_{11} on their own.
Going back to the calculated valence charges for the vanadium and oxygen sites given in table 4.9 we study whether the LDA findings point to charge ordering in one of the phases of V_6O_{11}. Different radii of the atomic spheres, necessary to fulfill the requirements of the atomic sphere approximation, prohibit a quantitative comparison of the valence charges but allow still for qualitative results. Concentrating on the high temparature modification we find roughly 0.2 electrons less for sites V1 and V3 than for V2 and V4. In contrast, the charges of the central sites V5 and V6 coincide. Note that the atomic spheres comprise about the same space for the pairs V1/V2, V3/V4, and V5/V6, respectively. In general, the calculated 2.5/2.7 V $3d$ valence electrons contradict the simple ionic picture, which predicts an average amount of 1.33 electrons per vanadium atom, see table 4.1. Instead of $V^{3.67+}$ ions we observe both $V^{2.3+}$ and $V^{2.5+}$ configurations, which is not surprising due to covalent V-O bonding. Ionic pictures naturally overestimate the charge transfer from the cationic vanadium to the anionic oxygen sites. Except for small variations the oxygen spheres comprise four $2p$ electrons. Further electrons enter the interstitial empty spheres and hence the open space between vanadium and oxygen atoms.
The accuracy of comparing calculated valences is of course limited since the assignment of electronic charge to specific atoms is arbitrary to some degree. Here we assign the charge located within an atomic sphere to the respective atomic site. Electrons entering interstitial empty spheres cannot be assigned to any specific atom. Importantly, differences in the occupation of equivalent states may be identified with much higher accuracy. Hence, in all probability, the $3d$ charge of both V1 and V3 is significantly reduced. Moreover, we obtain the same result for the low temperature phase, where all vanadium spheres reveal similar radii. The valence charge at the sites V1, V3, and V5 is about 0.2 electrons smaller than at the sites V2, V4, and V6, respectively. Together we can state a charge transfer of roughly 0.1 electrons per site from the 1-3-5 to the 2-4-6 chain – both above and below the transition. The different electron count in the two vanadium chains is related to their crystallographic inequivalence. While the electron donor atoms to some extend approach the VO_2 charge configuration, the acceptor chain resembles more the V_2O_3-type filling. The electronic configuration of VO_2 and V_2O_3 is apparently preferable to a mixture of

both. Note that the (sesquioxide-like) chain end sites do not generally gain excess charge at the expense of the (dioxide-like) chain centers. Hence it is not the electron count of a specific site but its local coordination which determines the electronic properties and the behaviour at the MIT. The calculated valence charges contradict ordered arrangements of V^{3+} and V^{4+} ions, which an ionic picture would predict. Although two kinds of vanadium valences in principle are confirmed, our results point to smaller charge differences. Deviations from the ionic picture are consistent with the susceptibility measurements by A. C. Gossard *et al.* [91], whereas we do not agree with the charge ordering proposed in this paper for the insulating phase.

The above findings have important implications for understanding V_2O_3 since the chain end atoms of the Magnéli phases resemble the local metal coordination of this material. According to the previous discussion the electronic properties of V_6O_{11} are influenced by essentially three effects. First, the center sites of the vanadium chains along c_{prut} reflect known characteristics of the embedded Peierls instability responsible for the MIT in VO_2. Second, the localization of the $d_{x^2-y^2}$ orbitals at the chain ends leaves them susceptible to electronic correlation. Third, the chain end d_{yz} states to roughly equal parts are subject to both metal-metal bonding within (along b_{prut}) and perpendicular to (along a_{prut}) the vanadium layers. As vanadium chains in V_2O_3 ($n = 2$) degenerate to pairs the dioxide-like Peierls instability does not contribute as a matter of principle. However, the localized nature of the $d_{x^2-y^2}$ orbitals and the importance of strong V-V coupling parallel to b_{prut} should be transferred to the sesquioxide. Due to the high crystal symmetry both effects could not be analyzed in previous studies of V_2O_3. As a consequence of the broken symmetry in the Magneĺi phases, combined with a closely related coordination of appropriate vanadium atoms, they are accessible to the present investigation of V_6O_{11}. The splitting of the a_{1g}-like states due to metal-metal overlap along c_{hex} might be less important than commonly assumed for V_2O_3. Moreover, the electronic structure and the MIT of V_6O_{11} remain a crucial test case for all theories aiming at a correct description of both VO_2 and V_2O_3. The MIT of V_6O_{11} arises as the combination of structural and electronic changes appearing either in the dioxide or in the sesquioxide.

4.4 Band Narrowing in V_4O_7

In the following, we continue with an investigation of V_4O_7. The band structure calculations are again based on the augmented spherical wave method. Effects of the sequioxide-like chain ends on the electronic properties can be analyzed in detail since the vanadium chains along c_{prut} contain only four atoms. Analogous with the preceding section we will discuss the electronic structure and its changes at the MIT in relation to the structural transformations occuring simultaneously. We will attain this goal because of the unified point of view we developed for the crystal structures of the Magnéli compounds including VO_2 and V_2O_3. Using the knowledge we established during the previous V_6O_{11} study it will be possible to group the calculated electronic bands into states behaving similar to the dioxide or the sesquioxide. Studying Magnéli phases characterized by different chain lengths enables us to understand the relations between the structural and electronic properties near the MITs since the materials can be analyzed on a common basis. The central

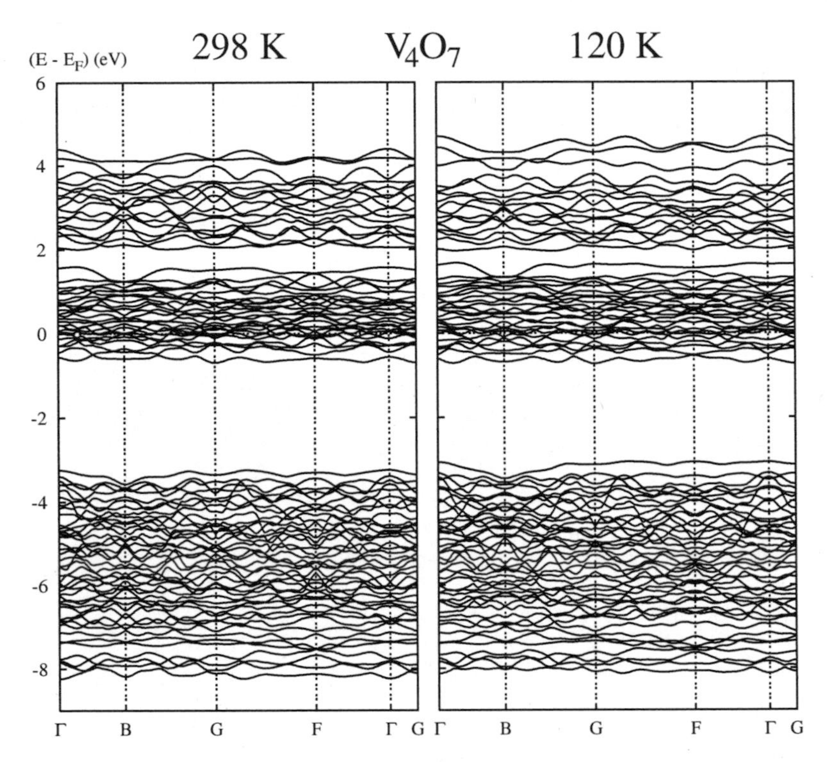

Figure 4.21: *Electronic bands of high as well as low temperature V_4O_7 shown along selected symmetry lines in the first Brillouin zone of the triclinic lattice as depicted in figure 4.16.*

results of these considerations have been published in a previous paper [104].
We begin the investigation of the electronic states arising from the LDA band structure calculation with an analysis of the gross features. For this purpose figure 4.21 displays the band structures for both the high and the low temperature V_4O_7 modification. In order to allow for an optimal comparison the same energy interval as used in the corresponding figures for VO_2, V_2O_3, and V_6O_{11} is chosen. Since the lattice symmetry of V_4O_7 resembles the triclinic arrangement of V_6O_{11} we apply the first Brillouin of the triclinic lattice, see figure 4.16. Even at first glance figures 4.15 and 4.21 show a fairly close relationship. In each case we observe three separated groups of electronic states. Due to a complex unit cell one finds a large number of bands, revealing dispersion throughout the first Brillouin zone, in small energy ranges. According to figure 4.21 the electronic states are affected by strong hybridization effects. Hence an investigation in terms of orbitally weighted band structures is not possible and we have to rely on the analysis of the local DOS to identify the electronic states which are relevant for the MIT of V_4O_7. The constitutional similarity in the crystal structures of the vanadium oxides VO_2, V_2O_3, V_4O_7, and V_6O_{11} explains

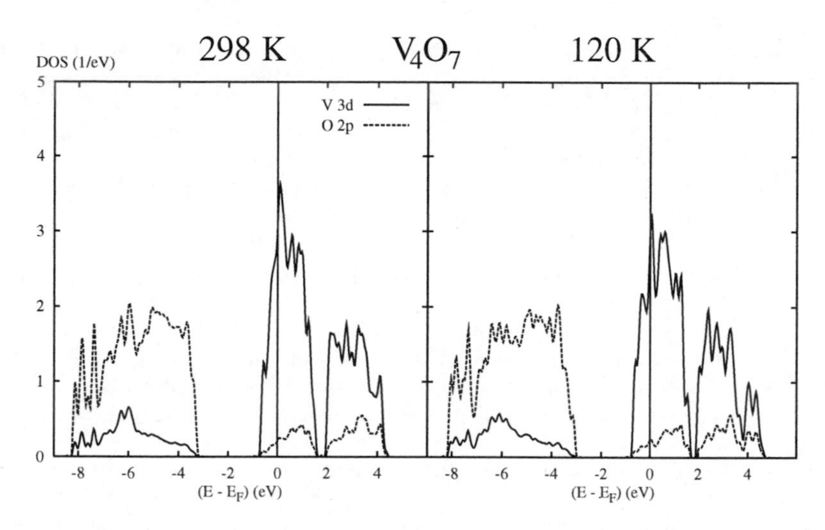

Figure 4.22: *Partial V 3d and O 2p densities of states (DOS) per vanadium atom resulting from the high (298 K) as well as low (120 K) temperature structure of the compound V_4O_7.*

the relations between their band structures, in particular the appearance of three groups of bands. Remembering the molecular orbital picture and the bonding-antibonding splitting due to overlap between V $3d$ and O $2p$ orbitals, we identify the bonding states with the energetically lowest group of bands in figure 4.21. At high temperatures they extend from -8.3 eV to -3.3 eV, whereas they shift to the region from -8.2 eV to -3.0 eV below the phase transition. The middle and highest group of bands correspond to the π^* and σ^* states, respectively. They are vanadium dominated and therefore separate into t_{2g} and e_g^σ contributions because of the octahedral coordination of the vanadium sites. While the t_{2g} group occupies the energy range from -0.7 eV to 1.6 eV in the metallic phase, its upper edge slightly shifts to 1.7 eV at low temperatures. Finally, the e_g^σ group starts at 2.0 eV in both phases but extends to somewhat higher energies in the insulating case, i.e. to 4.7 eV instead of 4.4 eV. For V_4O_7 the unit cell contains 8 vanadium and 14 oxygen atoms both at high and low temperatures. Therefore we have $8 \times 2 = 16$ vanadium $3d$ e_g^σ, $8 \times 3 = 24$ vanadium $3d$ t_{2g}, and $14 \times 3 = 42$ oxygen $2p$ bands in figure 4.21.
After this introductory survey of the band structures we now turn to the partial V $3d$ and O $2p$ densities of states given in figure 4.22, which can be compared one-to-one to figures 3.7 (VO_2), 3.13 (V_2O_3), and 4.17 (V_6O_{11}). In accordance with the molecular orbital picture, and confirming our earlier observations, the energetically lowest of the three distinct structures in the DOS predominantly shows O $2p$ character. In contrast, the remaining groups mainly trace back to V $3d$ states. Hybridization between these orbitals yields finite contributions of vanadium in the bonding and of oxygen in the antibonding region – very similar to those reported for V_6O_{11}. As expected, such hybridization effects are stronger in the case of σ-type V-O overlap. Because all the above results reflect almost completely

the findings of the preceding section we can integrate V_4O_7 and its MIT into the picture already developed. Because of a formal V $3d$ charge of 1.50 electrons per vanadium site, V_4O_7 systematically ranks between V_6O_{11} and V_2O_3. Thus the V $3d$ t_{2g} orbitals are filled to a somewhat larger extent than for V_6O_{11}, which is confirmed by figures 4.17 and 4.22. Although the calculation misses an energy gap for the experimental insulating phase, the LDA shortcomings do not affect the investigation of the relations between structural and electronic properties or the analysis of implications for the MIT of V_4O_7.
While the previous analysis of V_6O_{11} gave strong indications for the predominant influence of electron-lattice interaction in the dioxide-like and of electronic correlations in the sesquioxide-like regions of the crystal, the present work concentrates on the sesquioxide-related member V_4O_7, allowing conclusions about the MIT of V_2O_3. First remember the two different vanadium layers distinguished in the Magnéli phases. Separated by oxygen layers they alternate along a_{prut} and comprise the atoms V1/V3 or V2/V4 for V_4O_7 . By virtue of the alternating layers and due to relative shifts of the vanadium 4-chains along c_{prut} the end atoms V1 and V2 are located on top of each other. We will again refer the V $3d$ orbitals to local coordinate systems with the z-axis oriented parallel to the apical axis of the local octahedron and the x-axis parallel to c_{prut}, respectively. Using this reference frame the decomposition of the V $3d$ t_{2g} group of states into the symmetry components $d_{x^2-y^2}$, d_{yz}, and d_{xz} is shown in figures 4.23 and 4.24. The site-projected partial densities of states are given for the crystallographically inequivalent vanadium sites V1,V2,V3,V4. In the following, we concentrate on the findings for the 1-3 chains, which is motivated by similar modifications, in comparison to those known from V_2O_3, in the local surrounding of atom V1 at the MIT. According to the data denoted in figure 4.6 the V1-V2 and V1-V3 distances increase at the MIT – as do the distances a and b in the sesquioxide at the PM-AFI transition, see figure 3.10. Since in the latter compound all the vanadium atoms are crystallographically equivalent a complete symmetry analysis of the electronic states cannot be performed, which will be possible for V_4O_7. As demonstrated in the discussion of V_6O_{11}, the local atomic arrangements are directly related to the electronic structures of the respective sites. Thus the V1 site of V_4O_7 allows us to study the influence of the structural changes at the MIT of V_2O_3 in more detail than in a direct investigation.
The gross features of the DOS curves in figures 4.23 and 4.24 are very similar to those of VO_2, V_2O_3, and V_6O_{11}. As a consequence of V-O overlap and crystal field splitting at the vanadium sites we find two groups of antibonding states. More specifically, V $3d$ t_{2g} states occupy the region from roughly $-0.7\,\text{eV}$ to $1.7\,\text{eV}$. Furthermore, a finite t_{2g} DOS is found between $2.0\,\text{eV}$ and $4.7\,\text{eV}$ where the e_g^σ states dominate. Because the magnitude of the latter contributions equals the findings for V_6O_{11} the deviations from the ideal octahedral coordination of the vanadium sites is similar in both compounds and particularly similar to the sesquioxide. Oxygen contributions in the energy range of figures 4.23 and 4.24 are less than 10% of the total DOS at the Fermi energy. Both vanadium and oxygen contributions in the energy region where the respective other states dominate trace back to covalent V-O bonding. Obviously, the LDA calculation for the low temperature atomic arrangement completely misses an insulating band gap.
Changes in the V_4O_7 crystal structure at 250 K are mainly concerned with strong dimerization evolving in both the 1-3 and 2-4 metal chain, see figure 4.6. While the V3-V3 and V2-V4 bond lengths shorten from 2.79 Å and 2.93 Å to 2.67 Å and 2.83 Å, the remaining

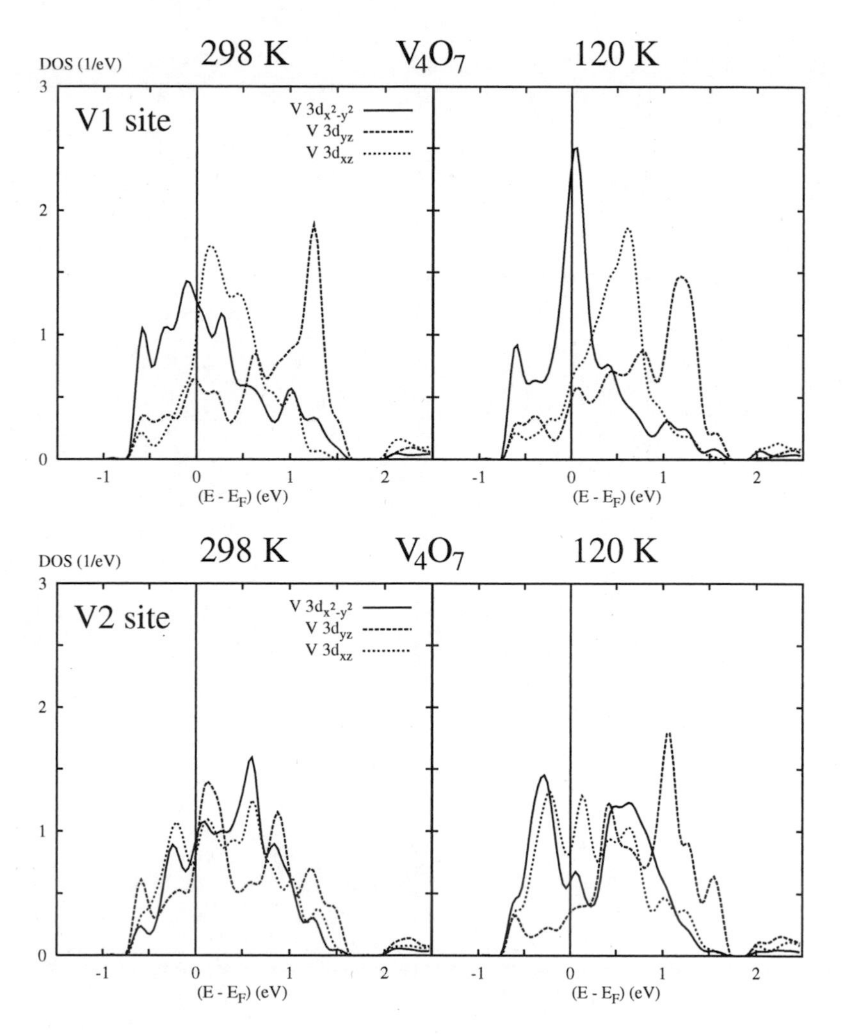

Figure 4.23: *Site-projected partial V 3d t_{2g} densities of states (DOS) per metal atom for the high and the low temperature crystal structure of V_4O_7: sites V1 and V2. The crystal structure is depicted in figure 4.6 and the orbitals refer to the local rotated reference frame.*

distances V1-V3 and V4-V4 elongate from 2.97 Å and 2.93 Å to 3.03 Å, respectively. As a consequence, we have an isolated V3-V3 vanadium pair in the first and two V2-V4 pairs in the second chain. In the 1-3 chain the end atoms decouple to a large degree from the central V3-V3 pair. Of course, the modifications resemble pairing effects associated with the

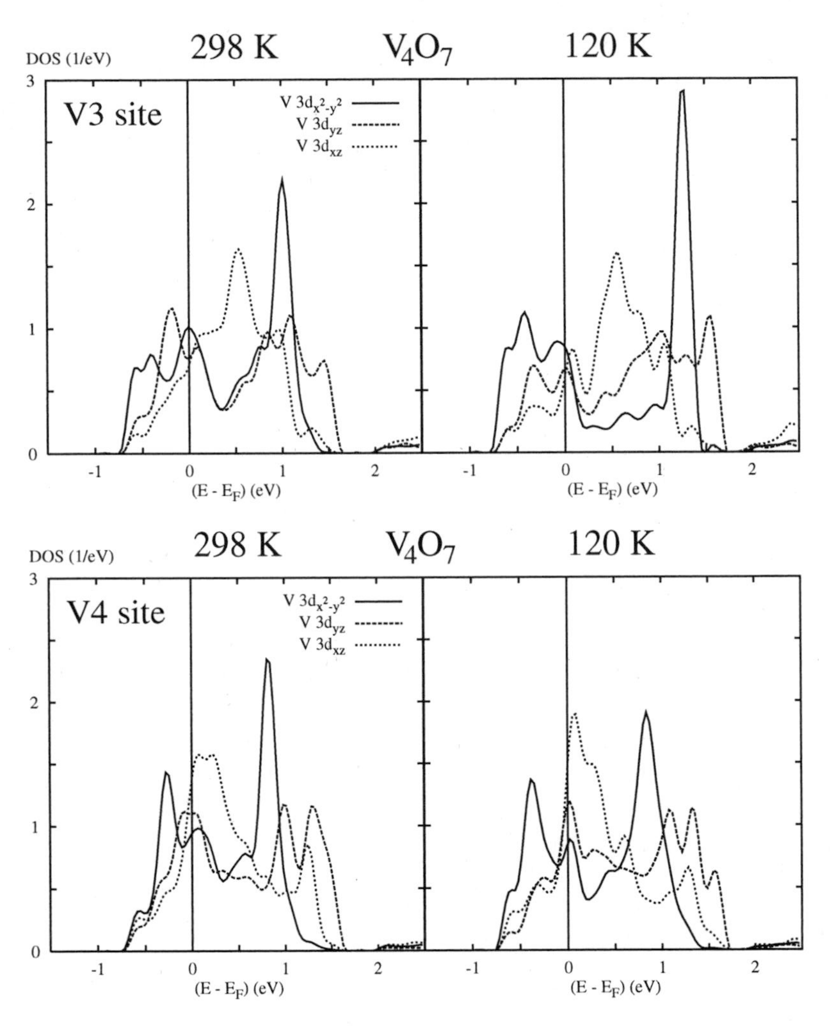

Figure 4.24: *Same representation as used in figure 4.23, but for the vanadium sites V3/V4.*

embedded Peierls instability of the dioxide. Nevertheless, the end atoms of the 1-3 chain likewise show structural changes known from the symmetry breaking PM-AFI transition of V_2O_3. More precisely, the distance of the V1 atom to its nearest vanadium neighbour along c_{prut} increases by 0.06 Å, which is about half the shift observed in the sesquioxide. Note that the dominating structural changes in V_2O_3 do not appear along a_{prut} but along c_{prut}. In contrast to the 1-3 chains, the end atoms of the 2-4 chains increase the coupling

to adjacent in-chain sites at low temperatures. They show the opposite trend than found in V_2O_3 and are therefore not suitable for studying the properties of the latter material. The different behaviour of the vanadium chains in V_4O_7 will be discussed later.
Beyond the metal-metal pairing also zigzag-type in-plane displacements of the vanadium atoms are observed in the case of V_4O_7. The distances between the vanadium sites and the octahedral centers amount to 0.31 Å/0.35 Å (V1), 0.32 Å/0.27 Å (V2), 0.15 Å/0.17 Å (V3), and 0.08 Å/0.09 Å (V4) in the high/low temperature modification. Apparently, the average bond length does not increase at the transition, which contrasts findings for VO_2 and V_6O_{11}, hence confirming the sesquioxide-like character of V_4O_7. Because of a slight rotation of the vanadium chains away from the c_{prut}-axis, the lateral displacement is much larger for the chain end sites V1 and V2. The rotation enables the latter sites to increase their distance along a_{prut}, which reflects the metal anti-dimerization of V_2O_3 parallel to c_{hex}. In general, the main part of the zigzag-type displacements is oriented perpendicular to c_{prut}. Atom V2 shifts almost parallel to the apical axis of its octahedron, i.e. parallel to the local z-axis, whereas atom V1 moves perpendicular to the local z-axis. However, due to the rotation of the local coordinate systems by 90° around the c_{prut}-axis, the displacements in all chains occur along the same direction. Indeed, on closer inspection we recognize the motions to be antiparallel, which increases the in-pair V1-V2 bond length. Together, the zigzag-type antiferroelectric mode of the 2-4 chain resembles the behaviour of the dioxide, whereas the 1-3 chain reveals distinct differences above as well as below the MIT. In retrospect, this interrelation applies also to V_6O_{11} where the displacements of the dioxide-like 1-3-5 chains point along the local z-axis rather than those of the 2-4-6 chains. However, the whole situation is less clear for V_6O_{11}.
Having a closer look at the V $3d$ t_{2g} partial DOS we find for the $d_{x^2-y^2}$ states a behaviour as observed in VO_2. Due to strong σ-type metal-metal bonding parallel to the vanadium chains the chain center atoms display a distinct splitting of the $d_{x^2-y^2}$ DOS. In the case of atom V3 a broad peak is observed from roughly -0.7 eV to 0.3 eV and a sharp peak at 1.0 eV. Because of the weaker V4-V4 bonding the splitting is smaller for the V4 site with peaks at approximately -0.3 eV and 0.8 eV. The chain end sites V1 and V2 are less affected by in-chain metal-metal bonding and consequently reveal a more compact $d_{x^2-y^2}$ DOS. At low temperatures the behaviour of both vanadium chains changes substantially. An increased V-V dimerization yields enhanced bonding-antibonding splitting of the V3 $d_{x^2-y^2}$ DOS. Additionally, by virtue of reduced V1-V3 overlap along c_{prut} the V1 $d_{x^2-y^2}$ states tend to localize as the interaction with the only intrachain bonding partner V3 is strongly reduced. Accordingly, the V1 $d_{x^2-y^2}$ DOS sharpens considerably and develops a pronounced peak close to the Fermi energy in the low temperature configuration. While the V4-V4 distance grows by 0.1 Å below the transition, the coupling of the V4 atoms to the chain end sites increases. The reduced V2-V4 bond length causes a stronger bonding-antibonding splitting of the V2 $d_{x^2-y^2}$ DOS, which is observable in figure 4.23.
The changes of the vanadium d_{xz} partial DOS at the phase transition are less significant. They consist mainly of an energetical up and downshift of the states as observed for the sites V1 and V2, respectively. Such shifts are due to increased and decreased vanadium displacements perpendicular to c_{prut}. Moving vanadium atoms laterally off the centers of gravity of their oxygen octahedra modifies the overlap between V $3d$ and O $2p$ orbitals. Note that the oxygen sublattice itself is hardly affected by the structural transformations.

Consequently, the shifts of the d_{xz} states are well understood in terms of the zigzag-type antiferroelectric displacements of the vanadium chains. Because these shifts are reduced for the chain center sites V3 and V4, see the former discussion, their influences are barely visible in the DOS. In the prototypical case of VO_2 lateral displacements are found only in the low temperature monoclinic phase, whereas V_4O_7 displays them similar at all temperatures. Generally, metal-metal bonding is less important for the d_{xz} orbitals. Thus no splitting in bonding and antibonding branches appears.
The situation is more complicated for the vanadium d_{yz} orbitals, which are considerably influenced by different kinds of hybridization. Recall from our earlier investigations that these electrons can be involved in metal-metal overlap across octahedral faces both along a_{prut} and b_{prut}. Furthermore, antiferroelectric-like displacements of the vanadium atoms lead to increased overlap between V $3d$ and O $2p$ states. This causes an energetical upshift of the antibonding vanadium dominated states. In the case of V_4O_7 the atoms V3 and V4 display neither relevant antiferroelectric shifts nor V-V interaction parallel to the a_{prut}-axis. There are simply no nearest vanadium neighbours along this direction. As a consequence, the pronounced double peak structure of the V3/V4 d_{yz} DOS is indicative of bonding-antibonding splitting due to V1-V3 and V2-V4 overlap, respectively. This is true for both the high and low temperature configuration. In analogy to the finite metal 3-chains of V_6O_{11} along b_{prut} we are confronted with vanadium pairs giving rise to V1-V3 and V2-V4 bonding across shared octahedral faces. In contrast to the central vanadium sites the chain end atoms V1 and V2, apart from participating in V1-V3/V2-V4 overlap parallel to b_{prut}, take part in the V1-V2 bonding along a_{prut}. The partial V1 and V2 d_{yz} densities of states hence consist of features reminiscent of the sites V2/V3 and V1/V4, respectively. Being subject to two different types of metal-metal overlap, the d_{yz} orbitals located at the chain ends undergo two intertwining bonding-antibonding splittings, which result in complicated DOS shapes. Here the curves reveal triangular shapes, especially in the low temperture phase, and hence to some extent resemble the shape of the V $3d$ a_{1g} DOS of V_2O_3, see figure 3.14. In addition, the form is affected by the discussed zigzag-type displacements of the chain end atoms, which intensify the metal-oxygen interaction. As a consequence, the center of gravity of the V1/V2 d_{yz} DOS shifts to higher energies below the MIT. The changes in the shape of the V2 d_{yz} DOS yield a closer resemblance to the V1 d_{yz} DOS, which can be traced back to an improved coupling between the chain end atoms at low temperatures. As the relative displacements of the atoms V1 and V2 parallel to c_{prut} are reduced, the V1-V2 bond penetrates the octahedral faces along a_{prut} more favorably and the overlap becomes more effective.
In conclusion, the electronic properties of V_4O_7 are strongly influenced by the local environments of the vanadium atoms, as already reported for V_6O_{11}. Generally, the overlap of the V $3d$ and O $2p$ orbitals places the V $3d$ t_{2g} states close to the Fermi energy. The detailed electronic structure is subject to the local metal-metal coordination. In particular, one can recognize a distinct bonding-antibonding splitting of the $d_{x^2-y^2}$ states at the chain center sites. In the low temperature phase the same applies to atom V2. All these splittings are analogous with the corresponding orbitals in VO_2 as well as V_6O_{11}, where they likewise are enforced by V-V dimerization. In contrast, the V1 $d_{x^2-y^2}$ DOS reveals a compact structure and a distinct sharpening below the transition because these states hardly take part in the dimerization along the vanadium chains. They may therefore be

susceptible to electronic correlations. For the d_{xz} and d_{yz} orbitals we obtain a noticeable response to the displacements of the vanadium atoms perpendicular to c_{prut}. Because the atoms move relative to the centers of their oxygen octahedra we end up with energetical up and downshifts in the DOS. However, for every atom the d_{yz} partial DOS is affected mainly by the V-V bonding across the shared octahedral faces. The chain center sites are subject to bonding only parallel to b_{prut} and reveal a characteristical two peak structure. Instead, atoms V1 and V2 additionally take part in d_{yz}-type overlap along a_{prut}, which results in densities of states similar to the a_{1g} DOS of V_2O_3. Remember that the latter mediates metal-metal bonding mainly parallel to $c_{\text{hex}} = a_{\text{prut}}$.
Similar to the V_6O_{11} findings, the LDA results indicate a small charge transfer between the two inequivalent vanadium chains. Recall the discussed limitations of assigning charge to specific atomic sites. Due to table 4.11 we have in the high temperature modification of V_4O_7 approximately 0.2 electrons less at the V2 than at the V1 site, whereas V3 and V4 reveal similar values. Note that in the pairs V1/V2 and V3/V4 the sphere radii are almost equal. The interpretation of the low temperature results is a little more complicated. However, taking into account the various sphere radii we can state reduced charges for both V2 and V4. The charge deficit may again be estimated to roughly 0.2 electrons per atom. Of course, the calculated 2.6-2.8 V $3d$ valence electrons do not agree with the simple ionic picture predicting an average amount of 1.50 electrons per vanadium atom, but due to covalent V-O bonding the difference is not surprising. In comparison to V_6O_{11} the atomic spheres of V_4O_7 comprise slightly more charge, which reflects the position of the material in the Magnéli series. We find no significant differences in the charges of the particular oxygen spheres since each of them contains roughly four $2p$ electrons. Further electrons enter the interstitial empty spheres and thus are difficult to classify in vanadium and oxygen-type contributions. Altogether we observe a charge transfer of about 0.1 electrons per site from the 2-4 to the 1-3 chain. In the high temperature phase only the chain end sites are involved in the transfer. As in the case of V_6O_{11} the different electron count is related to the crystallographical inequivalence of the two vanadium chains. While the 2-4 donator chain becomes slightly more VO_2-like, the 1-3 acceptor chain approaches the V_2O_3 configuration to some extent. The above LDA results qualitatively confirm x-ray results by M. Marezio *et al.* [88], who applied an ionic picture and thus reported chains of V^{3+} and V^{4+} ions running parallel to c_{prut}. Even though the ordered arrangement of varying valence charges is confirmed by the calculation, the charge discrepancy is much smaller due to covalent bonding. Nuclear magnetic resonance findings by A. C. Gossard *et al.* [87] support chains of differently charged atoms along c_{prut} but indicate incomplete differentiation into 3+ and 4+ valences. The charge ordering slightly increases below the MIT, which is confirmed by the LDA results.
In conlusion, similar to the situation in other Magnéli phases, particularly in V_6O_{11}, the electronic structure of V_4O_7 arises as a mixture of features characteristic of either vanadium dioxide or sesquioxide. The local environment of the chain end atom V1 bears close resemblance to the V_2O_3 structure since it is subject to an increased separation from its vanadium neighbours at the MIT. We are therefore able to draw important conclusions for V_2O_3. First, metal-metal bonding across octahedral faces parallel to a_{prut} seems to have the major effect on the shape of the d_{yz} DOS of V_4O_7 and thus on the shape of the a_{1g} DOS of V_2O_3. However, also the V-V hybridization across the faces parallel to b_{prut} is

large since it leads to strong splitting of the electronic states. Second, due to the reduced metal-metal interaction across the octahedral edges parallel to c_{prut} the V $d_{x^2-y^2}$ orbitals are substantially localized. Moreover, they undergo further localization at the transition. Both these facts should likewise apply to V_2O_3. This observation supports previous work by P. D. Dernier [64], who concluded from a comparison of pure as well as doped V_2O_3 and of Cr_2O_3 that the metallic properties of the sesquioxide are intimately connected to the V-V hybridization across the shared octahedral edges rather than with hopping processes within the vanadium pairs along c_{hex}. From this point of view the MIT arises from the crystal structure changes occuring at the transition. Via strong electron-phonon coupling the structural distortions translate into a narrowing of the bands perpendicular to c_{hex}. Remember that the decoupling of the vanadium atoms in the c_{prut}-direction is even stronger for the sesquioxide than for V_4O_7 because the corresponding bond length shortens by 0.12 Å (instead of 0.06 Å). The band narrowing eventually allows for an increased influence of electronic correlations in the insulating configuration. So far our treatment of V_2O_3 concerned only the symmetry breaking phase transition from the PM into the AFI phase, where we find substantial modification of the interatomic V-V spacings. However, P. Pfalzer *et al.* [67] reported on the discovery of local symmetry breaking in the PI phase of aluminium doped V_2O_3 giving rise to identical structural and electronic properties of the AFI and PI phase – at least on a local scale. Due to local distortions in the PI phase one may expect the same MIT scenario for both the PM-AFI and the PM-PI transition. In general, the in-plane band narrowing seems to play a key role for the MIT of V_2O_3.

4.5 Systematic Aspects of the Magnéli Series

Up to now we have analyzed the MITs of two special Magnéli phases: V_4O_7 and V_6O_{11}. Doing so has provided an explicit understanding of the local electronic properties of both dioxide and sesquioxide-like coordinated vanadium sites and their relevance for the phase transition. In particular, we have concentrated our efforts on the chain end sites in order to gain new insight into the mechanisms underlying the MITs of V_2O_3. The chain center sites have been understood in the framework of the embedded Peierls instability initially known from VO_2. In the following, we complement the above considerations by exploiting the structural systematics inherent in the Magnéli phases. Our particular interest regards the influence of the length of the vanadium chains on the electronic features. Varying the chain length systematically modifies the ratio of dioxide and sesquioxide-like sites. Especially a comparison of V_7O_{13} and V_8O_{15} in this context is useful since these materials are structurally most similar. Nevertheless, the former compound is the only member of the Magnéli series which does not undergo an MIT as a function of temperature.
In a comparative analysis of the Magnéli compounds the end members VO_2 and V_2O_3 take an exceptional postition since they either miss chain end or chain center vanadium sites. Additionally, these materials reveal a much higher symmetry of the crystal lattice than implied by the triclinic space group $P\bar{1}$ (C_i^1). The same is true for the monoclinic lattice of V_3O_5. Although the latter compound is even closer related to the sesquioxide than V_4O_7 it is nevertheless unsuitable for comparing to V_2O_3. Both V_5O_9 and V_8O_{15} are characterized by a superstructure doubling the volume of the unit cell in the insulating

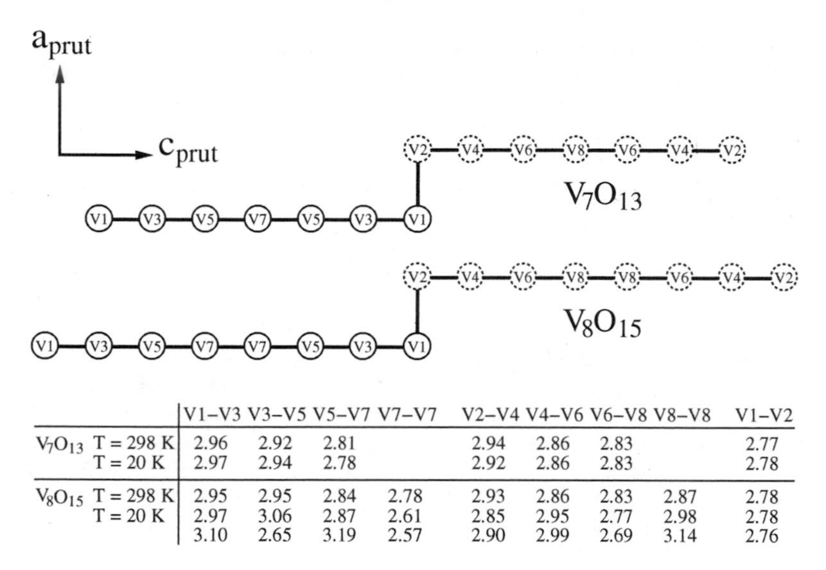

		V1–V3	V3–V5	V5–V7	V7–V7	V2–V4	V4–V6	V6–V8	V8–V8	V1–V2
V_7O_{13}	T = 298 K	2.96	2.92	2.81		2.94	2.86	2.83		2.77
	T = 20 K	2.97	2.94	2.78		2.92	2.86	2.83		2.78
V_8O_{15}	T = 298 K	2.95	2.95	2.84	2.78	2.93	2.86	2.83	2.87	2.78
	T = 20 K	2.97	3.06	2.87	2.61	2.85	2.95	2.77	2.98	2.78
		3.10	2.65	3.19	2.57	2.90	2.99	2.69	3.14	2.76

Figure 4.25: *Notation of the vanadium sites in V_7O_{13} and V_8O_{15}, where we observe metal chains of length $n = 7/n = 8$ along c_{prut}. In both cases eight inequivalent vanadium sites give rise to two types of chains. The spacial arrangement of the chains is similar to that of V_4O_7 and V_6O_{11}, see figures 4.6 and 4.7. The table gives measured V-V distances (in Å) for both the metallic (298 K) and the insulating (20 K) phase of V_7O_{13}/V_8O_{15} [96]. Below the transition the latter compound develops a superstructure with four inequivalent chains.*

phase. While the vanadium chains in the low temperature V_5O_9 configuration [105, 106] lose their inversion symmetry, one recognizes four kinds of inequivalent chains in the case of V_8O_{15}. The notation of the vanadium sites in V_7O_{13} and V_8O_{15} is given in figure 4.25. In the usual way we denote the inequivalent chains of the $n = 7$ compound V1-V3-V5-V7-V5-V3-V1 and V2-V4-V6-V8-V6-V4-V2. For V_8O_{15} the innermost site is doubled, which leads to the chains V1-V3-V5-V7-V7-V5-V3-V1 and V2-V4-V6-V8-V8-V6-V4-V2. Extra sites in the insulating superstructure are indicated by tildes. Similar to V_4O_7 and V_6O_{11} the oxygen sublattice of the $n = 8$ compound hardly changes at the MIT. The dominating structural modifications are again concerned with the V-V dimerization along c_{prut}, see the bond lengths summarized figure 4.25.
The investigations of V_7O_{13} and V_8O_{15} are based on similar LDA band structure calculations as used for the previous Magnéli compounds. Again the augmented spherical wave method is applied. In the partial DOS (not shown) we find three groups of bands, which correspond to the O $2p$, V $3d$ t_{2g}, and V $3d$ e_g^σ states. Due to a reduced formal electron count of 1.29 and 1.25 electrons per vanadium site for the $n = 7$ and $n = 8$ compound, respectively, the filling of the t_{2g} group is slightly smaller than for V_4O_7 and V_6O_{11}, see figures 4.22 and 4.17. Moreover, site-projected partial densities of states for V_7O_{13} and

V_8O_{15} reveal the typical features discovered before. In particular, the shapes of the DOS curves are completely understood in terms of the local coordination of the corresponding vanadium atoms, which seems to be true for all the Magnéli compounds. In the following discussion of V_7O_{13} and V_8O_{15} it is therefore useful to concentrate on two representative atomic sites: the central site V7 and the last but one site V3 of the 1-3-5-7 chain.
Figure 4.26 gives site-projected partial densities of states for the vanadium atoms V3 and V7 arising from the room temperature crystal structures of V_7O_{13} and V_8O_{15}. As usual, the representation of the V $3d$ symmetry components refers to the local rotated reference frame with the z and x-axis parallel to the apical axis of the local octahedron and to the c_{prut}-axis, respectively. V-O overlap combined with crystal field splitting ends up in two groups of antibonding states. While the t_{2g} interval reaches from approximately $-0.6\,\text{eV}$ to $1.8\,\text{eV}$, we observe additional t_{2g} contributions at energies above $2.0\,\text{eV}$, where the e_g^σ states dominate. Consistent with the positions of V_7O_{13} and V_8O_{15} in the Magnéli series, the latter are small – thus resembling the behaviour of VO_2 rather than of V_2O_3. Contributions due to O $2p$ orbitals, which arise from covalent V-O bonding, in the energy range used in figure 4.26 comprise less than 10% of the total DOS at the Fermi energy. Hence they are not included in the picture. When comparing the in-chain V-V distances in the high temperature modifications of V_7O_{13} and V_8O_{15} we find a very good agreement, see figure 4.25. Of course, we are confronted with additional V7-V7 and V8-V8 bonds in the $n = 8$ case, which are not available in chains of length $n = 7$. The oxygen sublattice of the compounds is almost identical, which likewise applies to the antiferroelectric-like lateral displacements of the vanadium atoms perpendicular to c_{prut} and the rotation of the metal chains off the c_{prut}-axis. Compared to V_4O_7 and V_6O_{11} the latter effects are weaker due to the dioxide-like nature of the long chain compounds. Bearing in mind the importance of the local atomic coordination for the DOS shape we expect a similar electronic structure for metallic V_7O_{13} and V_8O_{15}, which is confirmed by figure 4.26.
Consequently, the local $d_{x^2-y^2}$, d_{yz}, and d_{xz} densities of states resemble one another even with regard to the details of their shapes. This applies to the depicted sites V3 and V7 as well as to all other vanadium atoms. Due to strong σ-type metal-metal bonding along the c_{prut}-axis the chain center atoms are characterized by a distinct splitting of the $d_{x^2-y^2}$ DOS. In the case of V7 we hence find sharp peaks at about $-0.3\,\text{eV}$ and $1.1\,\text{eV}$. Because of the increasing V-V distances near the chain ends the bonding-antibonding splitting is noticeably smaller for the V3 $d_{x^2-y^2}$ DOS, where peaks occur at roughly $0.2\,\text{eV}$ and $0.7\,\text{eV}$. As a consequence of less efficient π-type V-V overlap via the d_{xz} orbitals the respective DOS is rather compact. In contrast, a substantial splitting in bonding and antibonding branches can be found in the d_{yz} DOS. For atoms V3 and V7 this shape traces back to metal-metal interaction along b_{prut}. The strength of the splitting is roughly identical for both sites. To conclude, the site-projected densities of states of high temperature V_7O_{13} and V_8O_{15} are understood analogous with our discussions in the preceding sections. By virtue of closely related V-V distances in both compounds similar DOS curves arise. It is therefore surprising that only one of the materials undergoes an MIT.
In order to illuminate the close relations between the electronic properties of V_7O_{13} and V_8O_{15} in more detail we consider the calculated valence charges summarized in table 4.14. The difficulty of assigning charge to specific atomic sites earlier was discussed at length. While the charges in the 1-3-5-7 chain of V_7O_{13} and the 2-4-6-8 chain of V_8O_{15} are more or

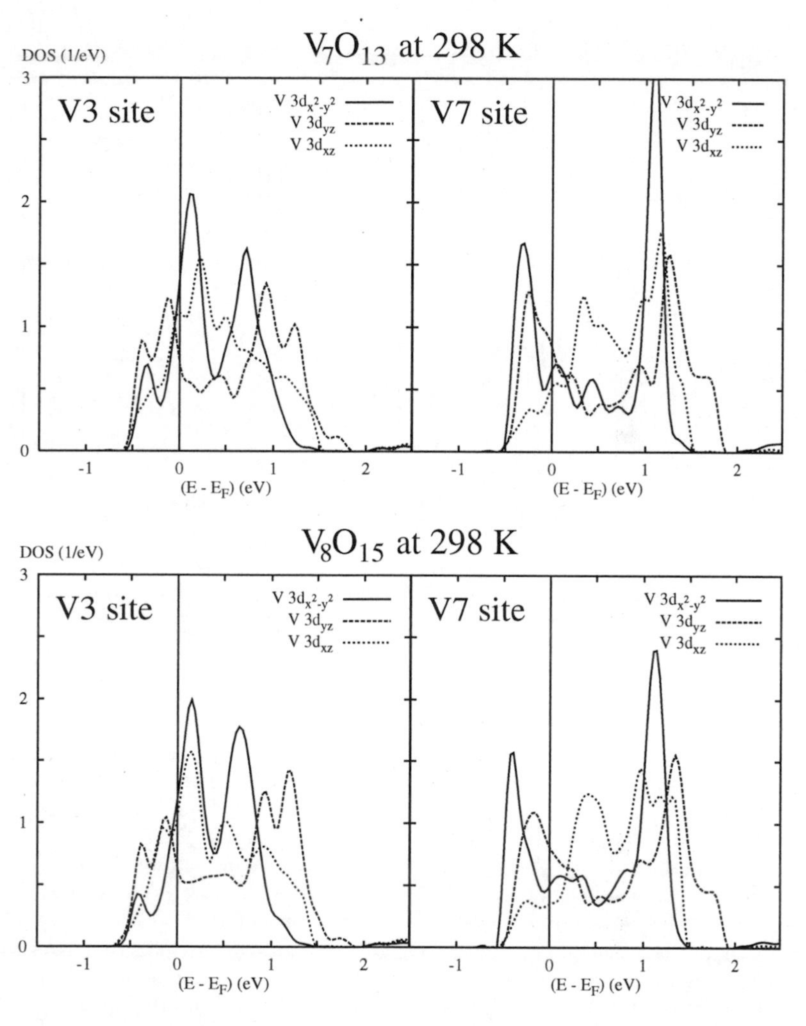

Figure 4.26: *Site-projected partial V 3d t_{2g} densities of states (DOS) per metal atom for high temperature V_7O_{13} and V_8O_{15}. The orbitals refer to the local rotated reference frame. A comparison of the DOS at the sites V3 and V7 (see figure 4.25) shows little discrepancy.*

less equally distributed among the vanadium sites, the other chains show a small shift of charge towards the chain centers, but charge transfer effects are small. Moreover, the electronic structure near the Fermi energy is determined by the local vanadium coordination. In the LDA calculation the metallic phases of V_7O_{13} and V_8O_{15} are barely distinguishable

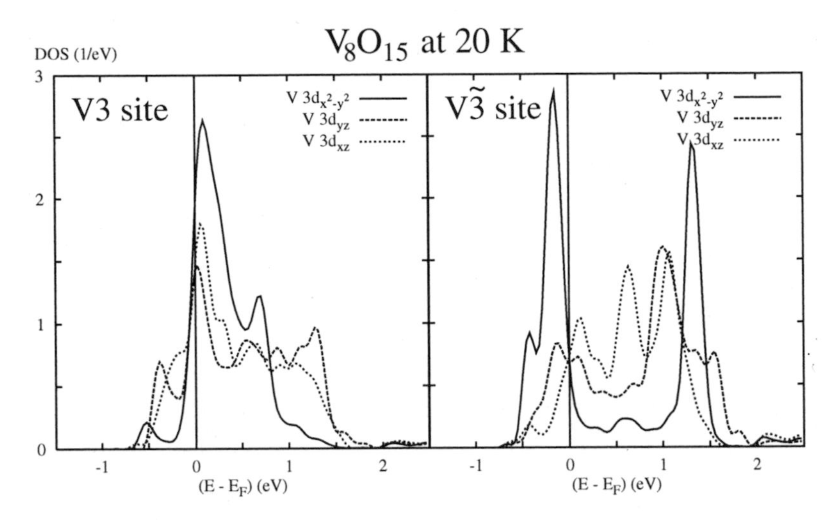

Figure 4.27: *Site-projected partial V 3d t_{2g} densities of states (DOS) per metal atom for sites V3 and V$\tilde{3}$ of low temperature V_8O_{15}. In the high temperature phase these sites are crystallographically equivalent. The orbitals again refer to the local rotated reference frame.*

as far as their local electronic characteristics are concerned. This points to the relevance of correlation effects for the electronic structures of the vanadium Magnéli phases.

An investigation of the low temperature V_8O_{15} modification reveals the same mechanisms as found previously in the Magnéli series. Depending on the bond lengths to the adjacent vanadium atoms along c_{prut} one finds either increased splitting or enhanced localization of the $d_{x^2-y^2}$ orbitals. The V-V bond lengths summarized in figure 4.25 indicate strong V-V dimerizations in the $\tilde{1}$-$\tilde{3}$-$\tilde{5}$-$\tilde{7}$, 2-4-6-8, and $\tilde{2}$-$\tilde{4}$-$\tilde{6}$-$\tilde{8}$ chains. In contrast, for the 1-3-5-7 chain no sequence of short and long bonds is found. While the innermost V-V distances of both the 2-4-6-8 and the $\tilde{2}$-$\tilde{4}$-$\tilde{6}$-$\tilde{8}$ chain elongate at the MIT from 2.87 Å to 2.98 Å and 3.14 Å, they shorten for the other chains from 2.78 Å to 2.61 Å and 2.57 Å. Thus the dimerization patterns are inversed. As a consequence, in the low temperature structure all vanadium sites, except for V1, V3, V5, and V$\tilde{1}$, are involved in strong V-V overlap along c_{prut}. The increased interaction leads to a stronger bonding-antibonding splitting of the $d_{x^2-y^2}$ DOS. For atoms affected by V-V dimerization one additionally obtains energetical upshifts of the d_{yz} and d_{xz} densities of states. In accordance with earlier findings they reveal typical changes in the electronic structure known from the MIT of VO_2. Although the energetical separation between the $d_{x^2-y^2}$ band and the remaining e_g^{π} states of V_8O_{15} increases in the low temperature phase it is far from complete.

The described Peierls-like situation does not account for the atoms V1, V3, V5, and V$\tilde{1}$, which are characterized by a decreasing V-V overlap below the MIT. Due to the reduced metal-metal bonding the $d_{x^2-y^2}$ densities of states sharpen considerably and in each case develop a pronounced peak close to the Fermi energy. Because of their sesquioxide-like co-

ordination the behaviour is not surprising for atoms V1 and $\tilde{V1}$. However, the separation from neighbouring vanadium atoms forces V3 and V5 to display the same localization of the $d_{x^2-y^2}$ orbitals. Hence electronic correlations are as important for these states as for the V_2O_3-type chain end sites. The question why one of the four vanadium chains in the low temperature superstructure of V_8O_{15} prefers localization to dimerization is difficult to answer. The calculated V 3d valence charges from table 4.14 do not give rise to any systematics. Note the increased charge of atom $\tilde{V3}$ (2.72 electrons) compared to atom V3 (2.54 electrons). Partial t_{2g} densities of states are depicted in figure 4.27. For the dioxide-like $\tilde{V3}$ site we recognize both an increased splitting of the $d_{x^2-y^2}$ DOS and an energetical upshift of the d_{yz} and d_{xz} curves, see figure 4.26. In contrast, the structurally separated V3 site is characterized by a very compact $d_{x^2-y^2}$ DOS located almost completely above the Fermi level hence explaining the reduced number of valence electrons. In comparison to the sesquioxide-like V1 site of V_4O_7, see figure 4.23, the filling of the low temperature $d_{x^2-y^2}$ states is significantly smaller due to the reduced electron count. All these findings do not imply a modification of our MIT scenario for the Magnéli phases as a mixture of features typical of either vanadium dioxide or sesquioxide.
We are nonetheless left with the missing MIT in V_7O_{13}. From the purely structural point of view the only difference between V_7O_{13} and V_8O_{15} is the number of dioxide-like chain center atoms. The local environments and hence the electronic properties of the individual sites are very similar. Recall that the MITs of the vanadium Magnéli phases arise as combinations of an embedded Peierls instabilty inherent in the vanadium chains and the electronic correlations relevant at the associated chain end sites. Because the compounds with small vanadium chains are dominated by the chain end sites we may speculate that sesquioxide-like states drive the MIT. On the other hand, the Peierls mechanism should be dominating for compounds with long chains. Going back to figure 4.2 the assumption of different driving forces for the MITs is consistent with the behaviour of the transition temperatures in the Magnéli series. With only one exception ($n = 5$) these temperatures monotonously decrease from V_3O_5 to V_7O_{13} where no longer a transition appears. In the case of V_5O_9 the peculiar superstructure evolving below the MIT may cause the observed deviations. Translated in our picture this kind of systematics corresponds to the decreasing influence of chain end atoms, in comparison to the growing chain length. The fewer atoms affected by V_2O_3-like states, the less advantageous a transition into an insulating phase. While the sesquioxide-like influences diminish with growing vanadium chains, the Peierls-like mechanism becomes stronger. The more chain center atoms available for the V-V dimerization, the bigger the advantage of forming the insulating modification. Bearing in mind the experimental transition temperatures these interrelations may point to a Peierls-driven MIT for V_8O_{15}, whereas for V_7O_{13} no mechanism seems strong enough to induce a transition. This speculative point of view is supported by pressure experiments of P. C. Canfield *et al.* [96], who found the phase transition of V_8O_{15} to disappear when hydrostatic pressure of more than 13 kbar is applied. We may presume that the pressure mainly influences the wide-stretching centers of the vanadium chains instead of the ends. Hence the magnitude of V-V dimerization seems essential for the MIT of V_8O_{15}. Unfortunately, up to now the crystal structure under pressure has not been determined, which prohibits an analysis by means of band structure calculations.

Chapter 5

Titanium Magnéli Phases

Analogous with the vanadium Magnéli phases the titanium oxides give rise to a homologous series of closely related compounds. The crystal structures of the titanium Magnéli phases, initially studied by S. Andersson and L. Jahnberg [83], are defined by the general stoichiometric formula

$$Ti_nO_{2n-1} = Ti_2O_3 + (n-2)TiO_2 \quad \text{where} \quad 3 \leq n \leq 10\,. \tag{5.1}$$

The above authors reported atomic arrangements similar to those discussed in detail for the vanadium case. In particular, the titanium Magnéli phases take an intermediary position between the rutile structure of TiO_2 ($n \to \infty$) and the corundum structure of Ti_2O_3 ($n = 2$). Additionally, the titanium Magnéli phases step by step transfer the formal metal valency stage three of the sesquioxide into the stage four of the dioxide. This is reflected by the systematics inherent in the crystal structures, which are normally viewed as rutile slabs of infinite extension separated by shear planes with a corundum-like atomic arrangement. While in the rutile regions the typical TiO_6 octahedra are coupled via edges, they share faces in the corundum shear planes.
To have a better understanding of the electronic properties of the titanium Magnéli compounds our representation of the crystal structures, developed for the isostructural vanadium Magnéli phases, is again suitable. For specific details see the discussions in section 4.1; here we give only a short review of the results. The crystal structures of the Magnéli phases are based on a regular three dimensional network consisting of oxygen octahedra. Apart from a slightly different buckling this network is identical for all members including TiO_2 and Ti_2O_3. Discrepancies between the compounds Ti_nO_{2n-1} arise from the filling of the oxygen octahedra with titanium atoms. Filled octahedra give rise to chains of length n parallel to c_{prut}, followed by $n-1$ empty sites. In the case of TiO_2 these titanium chains have infinite extension. Chains neighbouring along a_{prut} are shifted so that the end atoms exhibit one bonding partner in this direction, whereas the chain center sites reveal none. While octahedra neighbouring along a_{prut} and b_{prut} share faces, coupling between metal atoms in all other directions is via octahedral edges. Thus the chain end atoms are coordinated similar to Ti_2O_3, whereas the chain center sites reflect the atomic arrangement of TiO_2. This representation of the crystal structures makes it possible to refer all symmetry components of the Ti $3d$ orbitals to a common coordinate system with the z and x-axis

parallel to the apical axis of the local octahedron and the c_{prut}-axis, respectively.
Similar to the vanadium Magnéli phases the titanium systems show MITs as a function of temperature – accompanied by structural transformations. Despite homologous crystal structures and similar transition temperatures the temperature dependent conductivities in the titanium Magnéli series reveal only few systematic trends [107]. It is therefore assumed that different conductivity mechanisms have to be considered. Furthermore, some materials reveal several transitions of the metal-insulator or insulator-insulator type. Nevertheless, a comparison of the electronic properties of the vanadium and titanium based materials seems useful to arrive at a better understanding of the mechanisms underlying the transitions. Structural similarities allow for transferring some of our conclusions for the MITs of the vanadium Magnéli phases to the titanium systems. In particular, we will transfer our knowledge of V_4O_7 to the electronic structure of Ti_4O_7. These findings have been summarized in a recent paper [108].
Furthermore, a comparative analysis of the corundum based materials V_2O_3 and Ti_2O_3 will help to clarify the origin of the frequently discussed a_{1g} bands. The latter states play a central role in many theories aiming at an explanation of the MITs in V_2O_3. Understanding the relationship between crystal structures and electronic properties is essential for investigating the MITs. A comparison with the situation in Ti_2O_3, together with the study of an appropriate hypothetical crystal structure of vanadium sesquioxide, paves the way for interpreting the features of the V_2O_3 t_{2g} partial DOS.

5.1 Dimerization and Charge Transfer in Ti_4O_7

The Magnéli phase Ti_4O_7 has attracted considerable attention due to a peculiar electrical behaviour. At a temperature of about 154 K it reveals a metal-insulator transition and at 130 K an insulator-insulator transition with a thermal hysteresis of 12 K [109–111]. Both transitions are of first order and connected with a conductivity change of three orders of magnitude each. The magnetic susceptibility almost vanishes near 154 K, which has been interpreted as a change from Pauli paramagnetism to van Vleck behaviour, but displays hardly any change at 130 K [112]. The reduction of the magnetic susceptibility is induced by the formation of singlet Ti^{3+}-Ti^{3+} pairs – so-called bipolarons – giving rise to the MIT. While the bipolarons are disordered at higher temperatures, they order below 130 K thus yielding the second phase transition. Confirming the anisotropy of the crystal structure the electrical resistivity of Ti_4O_7 is very anisotropic. For pressures higher than 40 kbar the bipolaron formation is suppressed [113]. The crystal structures of Ti_4O_7 above 154 K and below 130 K have been determined by M. Marezio *et al.* [114, 115]. Relating the average Ti-O bond lengths due to the unequal TiO_6 octahedra below 130 K to an elementary ionic picture the authors assigned the valances 3+ and 4+ to the titanium sites, giving rise to non-magnetic Ti^{3+}-Ti^{3+} pairs. Between the transitions the crystal structure is related to the room temperature findings without titanium pairing and charge separation. Both low temperature phases were proposed to be insulating because of a localization of the Ti $3d$ states. For the intermediate phase Ti-Ti pairing without long-range order was suggested. In general, the structures of the titanium Magnéli compounds with parameters $n \geq 5$ are investigated less accurately, compared to Ti_4O_7 [107]. The electrical behaviour of Ti_5O_9

	High temperature structure			Low temperature structure		
Atom	x	y	z	x	y	z
Ti1	0.9880	0.9457	0.0977	0.9449	0.9077	0.1124
Ti2	0.9936	0.4490	0.0973	0.0230	0.4712	0.0913
Ti3	0.9673	0.9667	0.3744	0.0095	0.0046	0.3635
Ti4	0.9832	0.4835	0.3677	0.9758	0.4997	0.3687
O1	0.6905	0.4095	0.0524	0.6752	0.3927	0.0548
O2	0.3313	0.5619	0.1031	0.3285	0.5598	0.1043
O3	0.6837	0.4661	0.1715	0.7031	0.4803	0.1626
O4	0.3848	0.6336	0.2220	0.3848	0.6557	0.2156
O5	0.6813	0.4033	0.3372	0.6540	0.3684	0.3513
O6	0.3184	0.5814	0.3844	0.3277	0.6067	0.3760
O7	0.6691	0.4392	0.4695	0.6742	0.4447	0.4669

Table 5.1: *Atomic positions for the high and low temperature structure of Ti_4O_7 as used in the calculations. These data have been determined by Y. Le Page and M. Marezio [120] and the coordinates refer to the primitive translations proposed by H. Horiuchi et al. [98].*

is found to be similar to the case of Ti_4O_7, but there appear no significant changes in the crystal structure [116]. For Ti_6O_{11} a transition occurs at 150 K, which corresponds to an emerging superstructure with reduced Ti^{3+}-Ti^{3+} distances [117, 118].
Concerning the structural modifications accompanying the MIT, the titanium compound Ti_4O_7 behaves similar to V_4O_7 [88]. A large jump of the entropy at the 130 K transition was found by means of heat capacity measurements [110, 111], indicating strong influence of disorder. Therefore the intermediate phase of Ti_4O_7 often is interpreted as a partially ordered bipolaron liquid. In the vanadium doped system $Ti_{1-x}V_xO_7$ three phases appear up to $x < 0.35$ [119]. For $x > 0.35$ only one transition (into a disordered insulating state) is found. The interpretation in terms of an order-disorder transition at 130 K was called into question by the observation of a fivefold superstructure in the intermediate phase of Ti_4O_7, indicating long-range ordered titanium valences [120]. Hence a superstructure is inconsistent with the hypothesis of independently mobile bipolarons yielding the electrical conductivity. Local dynamical disorder of the Ti^{3+}-Ti^{3+} pairs may still occur.
Photoemission as well as O $1s$ x-ray absorption spectra of Ti_4O_7 have been reported by M. Abbate *et al.* [121]. Photoemission data obtained at 300 K show Ti $3d$ bands in the vicinity of the Fermi level and O $2p$ bands at lower energies from about -8 eV to -4 eV. In the insulating phase at 50 K the Ti $3d$ states shift about 0.25 eV towards the oxygen bands. X-ray absorption spectroscopy probes the unoccupied electronic states in the conduction band. Spectra taken at 300 K show two broad maxima arising from Ti $3d$ bands separated by 2.4 eV and Ti $4s/4p$ bands at higher energies. The double peak structure of the Ti $3d$ states (with t_{2g} and e_g subbands) arises from crystal field effects. It shifts roughly 0.45 eV to higher energies in the insulating phase at 80 K. While the absorption spectra change suddenly around 150 K, they are otherwise basically temperature independent. A recent photoemission study by K. Kobayashi *et al.* [122] tackles all three phases of Ti_4O_7. While the spectra of the low temperature insulating phase show a finite gap at the Fermi level,

the spectra of the high temperature insulating phase are gapless, which is interpreted as a soft Coulomb gap due to dynamical disorder. While the insulator-insulator transition at 130 K seems to be dominated by disorder effects, the origin of the 154 K MIT is not yet understood in detail. In particular, the driving force of the metal-metal pairing gives rise to open questions. Independent of possible superstructures between 154 K and 130 K this pairing seems to be essential for the insulating phases. In the following, we thus investigate the relations between the structural and electronic properties of Ti_4O_7 by means of LDA band structure calculations. Since only the influence of the titanium pairing is of interest here, not disorder effects, we compare the metallic to the ordered insulating phase.
Structural data for the band structure calculations of Ti_4O_7 are taken from Y. Le Page and M. Marezio [120]. Both the titanium and oxygen atoms are located at the Wyckoff positions (2i): $\pm(x, y, z)$. The triclinic lattice parameters are $a_L = 5.597$ Å, $b_L = 7.125$ Å, $c_L = 20.429$ Å, $\alpha_L = 67.70°$, $\beta_L = 57.16°$, and $\gamma_L = 108.76°$ at high temperatures (298 K). Below the phase transition (115 K) one finds $a_L = 5.626$ Å, $b_L = 7.202$ Å, $c_L = 20.260$ Å, $\alpha_L = 67.90°$, $\beta_L = 57.69°$, and $\gamma_L = 109.68°$. All these values refer to the unit cell choice of Y. Le Page and P. Strobel as denoted in equation (4.4). Inserting $n = 4$ in the latter relation yields primitive translations connected to the rutile lattice by

$$\begin{pmatrix} \mathbf{a}_L \\ \mathbf{b}_L \\ \mathbf{c}_L \end{pmatrix} = \begin{pmatrix} -1 & 0 & 1 \\ 1 & 1 & 1 \\ 0 & 0 & 7 \end{pmatrix} \begin{pmatrix} \mathbf{a}_R \\ \mathbf{b}_R \\ \mathbf{c}_R \end{pmatrix} . \tag{5.2}$$

The primitive translations of H. Horiuchi *et al.* take the form

$$\begin{pmatrix} \mathbf{a}_H \\ \mathbf{b}_H \\ \mathbf{c}_H \end{pmatrix} = \begin{pmatrix} -1 & 0 & 1 \\ 1 & 1 & 1 \\ 0 & \frac{7}{2} & \frac{7}{2} \end{pmatrix} \begin{pmatrix} \mathbf{a}_R \\ \mathbf{b}_R \\ \mathbf{c}_R \end{pmatrix} \tag{5.3}$$

and together with equation (5.2) we obtain the relation

$$\begin{pmatrix} \mathbf{a}_H \\ \mathbf{b}_H \\ \mathbf{c}_H \end{pmatrix} = \begin{pmatrix} 1 & 0 & 0 \\ 0 & 1 & 0 \\ \frac{7}{2} & \frac{7}{2} & -\frac{1}{2} \end{pmatrix} \begin{pmatrix} \mathbf{a}_L \\ \mathbf{b}_L \\ \mathbf{c}_L \end{pmatrix} . \tag{5.4}$$

Coordinates given in the L-system can directly be transformed into the H-system

$$\begin{pmatrix} x \\ y \\ z \end{pmatrix}_H = \begin{pmatrix} 1 & 0 & 7 \\ 0 & 1 & 7 \\ 0 & 0 & -2 \end{pmatrix} \begin{pmatrix} x \\ y \\ z \end{pmatrix}_L + \begin{pmatrix} 1/2 \\ 0 \\ 1/2 \end{pmatrix}_L , \tag{5.5}$$

where the additional translation vector must be added because of different choices of the origin in both coordinate systems. Applying equation (5.5), the positional parameters of Y. Le Page and M. Marezio for the high and low temperature Ti_4O_7 modification yield the structural input for the band structure calculations, see the atomic positions in table 5.1. Apart from four crystallographically inequivalent titanium sites we have to deal with seven oxygen classes. By virtue of the inversion symmetry of the crystal lattice the unit cell finally comprises 22 atoms.

	High temperature structure		Low temperature structure	
Atom	Radius	Charge	Radius	Charge
Ti1	2.2809	1.7590	2.3359	1.8604
Ti2	2.3096	1.7153	2.2051	1.5424
Ti3	2.3829	1.6475	2.4554	1.8235
Ti4	2.3739	1.6633	2.2188	1.5668
O1	2.0039	4.4381	1.9775	4.3582
O2	1.8843	4.0600	1.7921	3.9957
O3	2.0002	4.4037	1.9950	4.5375
O4	2.0507	4.4683	2.0290	4.3874
O5	1.8608	4.0350	1.9058	4.0467
O6	1.9441	4.1734	1.9399	4.1323
O7	1.9367	4.1465	1.8032	3.9683

Table 5.2: *Atomic sphere radii of titanium and oxygen (in a_B) as well as calculated LDA valence charges (Ti 3d or O 2p) for both the high and the low temperature phase of Ti_4O_7.*

Using the lattice parameters reported by Y. Le Page and M. Marezio (for the L-system), the triclinic unit cell of Ti_4O_7 in Cartesian coordinates is established analogous with the V_6O_{11} case, see equations (4.10) to (4.14). The resulting primitive translations belonging to the L-system are transformed via equation (5.4) thus giving rise to the unit cell of the H-system. For the high temperature Ti_4O_7 structure at 298 K we find ($A = 8.8789\,a_B$)

$$\mathbf{a}_H = A \begin{pmatrix} -1.0000 \\ 0.0000 \\ 0.6473 \end{pmatrix}, \quad \mathbf{b}_H = A \begin{pmatrix} 1.0000 \\ 0.9384 \\ 0.6473 \end{pmatrix}, \quad \mathbf{c}_H = A \begin{pmatrix} 0.0010 \\ 3.4458 \\ 2.3633 \end{pmatrix}. \tag{5.6}$$

Moreover, the low temperature structure at 115 K results in ($A = 8.9933\,a_B$)

$$\mathbf{a}_H = A \begin{pmatrix} -1.0000 \\ 0.0000 \\ 0.6305 \end{pmatrix}, \quad \mathbf{b}_H = A \begin{pmatrix} 1.0000 \\ 0.9448 \\ 0.6305 \end{pmatrix}, \quad \mathbf{c}_H = A \begin{pmatrix} 0.0054 \\ 3.4326 \\ 2.2888 \end{pmatrix}. \tag{5.7}$$

These primitive translations refer to Cartesian coordinates and are used with the atomic positions summarized in table 5.1.

Because of the openness of the Ti_4O_7 crystal structure empty spheres have to be applied in order to properly model the crystal potential. Both in the high and low temperature configuration it is sufficient to introduce 11 crystallographically inequivalent classes and place altogether 20 empty spheres in the unit cell. In doing so the linear overlap of physical spheres is kept below 19% and the overlap of any pair of physical and empty spheres below 24%. Finally, a unit cell comprises 42 spheres. The radii of the titanium and oxygen spheres are summarized in table 5.2. In addition, for later use the table contains the valence charges arising from the LDA band structure calculations. The basis sets taken into account in the secular matrix comprise Ti $4s$, $4p$, $3d$, $(4f)$ and O $2s$, $2p$, $(3d)$ orbitals as well as states due to the additional augmentation spheres. More precisely, the configurations of the latter reach from $1s$, $2p$, $(3d)$ to $1s$, $2p$, $3d$, $4f$, $(5g)$. As usual, the states

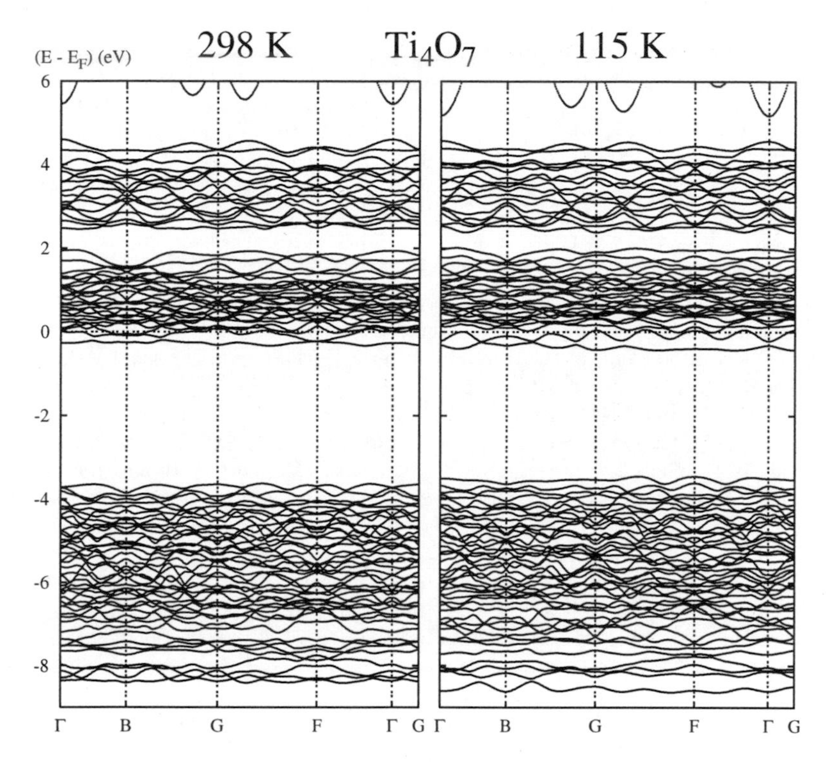

Figure 5.3: *Electronic bands of high and low temperature Ti_4O_7 displayed along selected symmetry lines in the first Brillouin zone of the triclinic lattice as depicted in figure 4.16.*

denoted in parentheses enter the calculation as tails of other orbitals. Convergence of the results with respect to the fineness of the **k**-space grid can be achieved by increasing the number of **k**-points taken into consideration in the Brillouin zone integration. For Ti_4O_7 this number increased from initially 108 to 256, 864, 2048 points in the irreducible wedge of the Brillouin zone. Self-consistency of the charge density was assumed for deviations of the atomic charges and the total energy of subsequent iterations less than 10^{-8} electrons and 10^{-8} Ryd, respectively. All the technical details apply to both the high and the low temperature phase.

The discussion of the LDA results for Ti_4O_7 starts with a very brief analysis of the band structure. Figure 5.3 shows the electronic bands resulting from both the high and the low temperature crystal structure along selected high symmetry lines in the triclinic Brillouin zone, see figure 4.16. Since our Ti_4O_7 investigation is based upon a comparison with the corresponding vanadium compound V_4O_7 the following results are presented in analogy to section 4.4. As expected from the close structural relationship connecting the vanadium and titanium Magnéli phases, figure 5.3 reflects the characteristics observed for V_4O_7. We

observe three well separated groups of bands, which extend from about −8.5 eV/−8.7 eV to −3.7 eV, from −0.5 eV to 2.0 eV, and from 2.3 eV to 4.7 eV. Compared to the vanadium system the energetical separation of the lowest and the middle group has grown by about 0.6 eV. This fact indicates a stronger bonding-antibonding splitting arising from the overlap of Ti $3d$ and O $2p$ orbitals. Applying the molecular orbital picture the lowest group of bands in figure 5.3 is identified with the bonding branch. The antibonding π^* and σ^* states correspond to the middle and the highest group, respectively. Due to the octahedral coordination of the metal sites the electronic states split up into contributions with e_g^σ and t_{2g} symmetry.
Compared to the V_4O_7 band structure, see figure 4.21, the middle (t_{2g}) group of bands in figure 5.3 reveals noticeably less states below the Fermi energy, which corresponds to the reduced electron count in the titanium system. In contrast to a formal V $3d$ valence charge of 1.50 electrons, here we have to deal with $(4 \cdot 4 - 7 \cdot 2)/4 = 0.50$ electrons per titanium site. As in V_4O_7 the Fermi energy falls into the t_{2g} group of bands. Both above and below the MIT the unit cell of Ti_4O_7 contains 8 titanium and 14 oxygen atoms. One thus finds 16 titanium $3d$ e_g^σ, 24 titanium $3d$ t_{2g}, and 42 oxygen $2p$ dominated bands in figure 5.3. The two lowest bands of the t_{2g} group effectively separate from the remaining bands at low temperatures. However, due to the LDA shortcomings the band structure calculation misses an insulating energy gap for the experimental insulating phase, which reflects the situation reported for VO_2. As in the case of the dioxide the LDA results still allow us to interpret characteristic features of the MIT of Ti_4O_7.
Similar to our procedure for the vanadium Magnéli phases we have to analyze the shapes of side-projected DOS curves in order to correlate the crystal structure modifications at the MIT of Ti_4O_7 with specific electronic changes. To this end we first turn to figure 5.4, which gives partial Ti $3d$ and O $2p$ densities of states. In accordance with the molecular orbital picture the lowest of the three distinct structures in the DOS is dominated by O $2p$ states, whereas the remaining groups predominately trace back to Ti $3d$ orbitals. The influence of hybridization between metal $3d$ and oxygen $2p$ orbitals is most similar in the cases of V_4O_7 and Ti_4O_7, see the contributions of vanadium/titanium in the bonding and of oxygen in the antibonding region. Of course, hybridization is stronger for σ-type orbital overlap. The energetical separation between the centers of gravity of the t_{2g} and the e_g^σ DOS amounts to about 2.7 eV and thus compares well with x-ray absortion experiments by M. Abbate *et al.* [121]. Moreover, the energetical position and width of the O $2p$ DOS agrees well with the photoemission data given by these authors.
In general the Ti $3d$ and O $2p$ densities of states are similar for the high and low temperature configuration. However, on closer inspection of the Ti $3d$ dominated states, striking alterations develop at the phase transition, particularly in the occupied t_{2g} range. This is clearly observable in figure 5.5, which displays the Ti $3d$ DOS decomposed into contributions from either the Ti1-Ti3-Ti3-Ti1 or the Ti2-Ti4-Ti4-Ti2 chain. Remember that the V_4O_7 crystal structure gives rise to alternating vanadium layers – mutually separated by layers of oxygen. The metal layers either comprise vanadium 4-chains of the type V1-V3-V3-V1 or of the type V2-V4-V4-V2. Corresponding titanium layers in the case of Ti_4O_7 consist either of Ti1-Ti3-Ti3-Ti1 or of Ti2-Ti4-Ti4-Ti2 chains. Therefore the schematic representation of V_4O_7 in figure 4.6 is useful to illustrate also the atomic arrangement in the metal sublattice of the titanium system. While for high temperature Ti_4O_7 the occu-

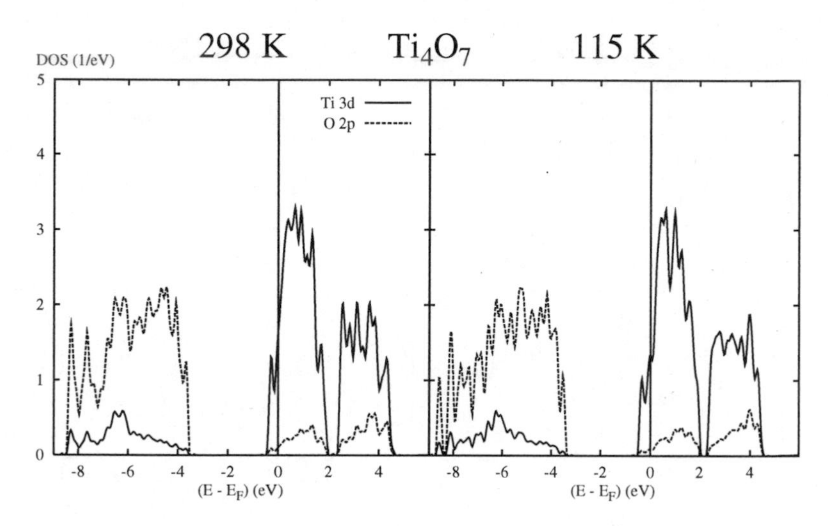

Figure 5.4: *Partial Ti 3d and O 2p densities of states (DOS) per titanium atom resulting from the high (298 K) as well as low (115 K) temperature structure of the compound Ti_4O_7.*

pied Ti $3d$ DOS in figure 5.5 is almost the same for both titanium chains, we find below the MIT strong increase and decrease of the contributions arising from the 1-3 and the 2-4 chain, respectively. This observation is indicative of considerable charge transfer. To be more specific, the charge difference between the chains amounts to roughly 0.9 electrons per formula unit. Moreover, states in the Ti $3d$ e_g^σ group of bands become rearranged at the phase transition. In particular, we find energetical down and upshift of the centers of gravity of these states for the 1-3 and 2-4 chain, respectively.
We attribute the charge transfer and the rearrangement of the e_g^σ states to differences in the average Ti-O bond lengths emerging in the low temperature phase [88]. While at room temperature this length is the same for all titanium sites, it increases for the 1-3 chain and decreases for the 2-4 chain below the MIT. As a consequence, $3d$ charge is transferred from Ti2/Ti4 to Ti1/Ti3, leaving the $3d$ states of the 2-4 chain almost unoccupied. Therefore the titanium configurations d^0 and d^1 characterize the inequivalent chains, contrasting the formal $d^{0.5}$ valence of the high temperature phase. In addition, the shrinking and inflating of the oxygen octahedra affects the bonding-antibonding splitting. Due to their σ-type Ti-O overlap this applies especially to the e_g^σ orbitals. A smaller/larger bonding-antibonding splitting in the 1-3/2-4 titanium chains leads to the observed downshift/upshift of the Ti $3d$ dominated antibonding states.
As for the vanadium system the high temperature Ti_4O_7 modification reveals rather little dimerization. Accordingly, the Ti-Ti distances along the chains amount to 3.02 Å (Ti1-Ti3), 2.89 Å (Ti3-Ti3), 3.02 Å (Ti2-Ti4), and 2.94 Å (Ti4-Ti4). Below the phase transition the respective values are 2.79 Å (Ti1-Ti3), 3.11 Å (Ti3-Ti3), 3.06 Å (Ti2-Ti4), and 3.01 Å (Ti4-Ti4). In contrast to the results for V_4O_7 only one of the titanium chains along c_{prut}

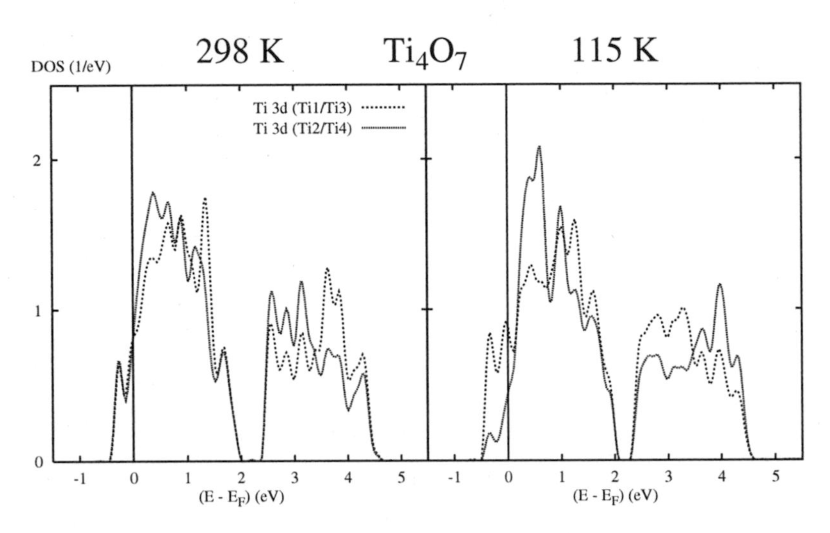

Figure 5.5: *Decomposition of the high and low temperature partial Ti 3d densities of states (DOS) of the compound Ti_4O_7 into contributions due to the chains Ti1/Ti3 and Ti2/Ti4.*

exhibits a strong Ti-Ti dimerization at low temperatures. Regarding the distortion patterns, the V2-V4 pair of V_4O_7 corresponds to the Ti1-T3 pair of Ti_4O_7 and the V1-V3 anti-pair to the Ti2-Ti4 anti-pair. Moreover, the length of the titanium 4-chains is larger than found in the vanadium system. The same is true for the metal-metal distance across shared octahedral faces along a_{prut}. For the Ti1-Ti2 bond length we find the values 2.81 Å and 2.86 Å in the metallic and insulating phase, respectively. In contrast, the V1-V2 distance hardly changes for V_4O_7. While isolated Ti1-Ti3 pairs characterize low temperature Ti_4O_7, the metal-metal coupling within the 2-4 chains decreases significantly at the MIT. Hence the 1-3 titanium chain resembles the pairing effects associated with an embedded Peierls instability, whereas the 2-4 chain is influenced by reduced Ti-Ti overlap and thus localization of the Ti 3*d* orbitals – particularly those oriented along the c_{prut}-axis.
In order to investigate the charge redistribution at the MIT of Ti_4O_7 in more detail one decomposes site-projected Ti 3*d* t_{2g} densities of states into their symmetry components. Again it is reasonable to refer these densities to local coordinate systems with the z-axis oriented parallel to the apical axis of the local octahedron and the x-axis parallel to c_{prut}, respectively. The results are depicted in figures 5.6 and 5.7. Obviously, the gross features of the DOS curves reflect the characteristics known from V_4O_7. Although the Ti 3*d* t_{2g} states occupy predominately the energy region from -0.4 eV to 2.0 eV, a finite t_{2g} DOS remains above 2.3 eV, i.e. in the range of the e_g^σ states. The latter contributions point to deviations from an ideal octahedral coordination of the titanium sites. Since octahedral deformations are similar in V_4O_7 and Ti_4O_7 the magnitude of the admixtures is roughly the same. Contributions of oxygen states in the energy range displayed in figures 5.6 and 5.7 amount to less than 10% of the total DOS at the Fermi energy but are still indicative

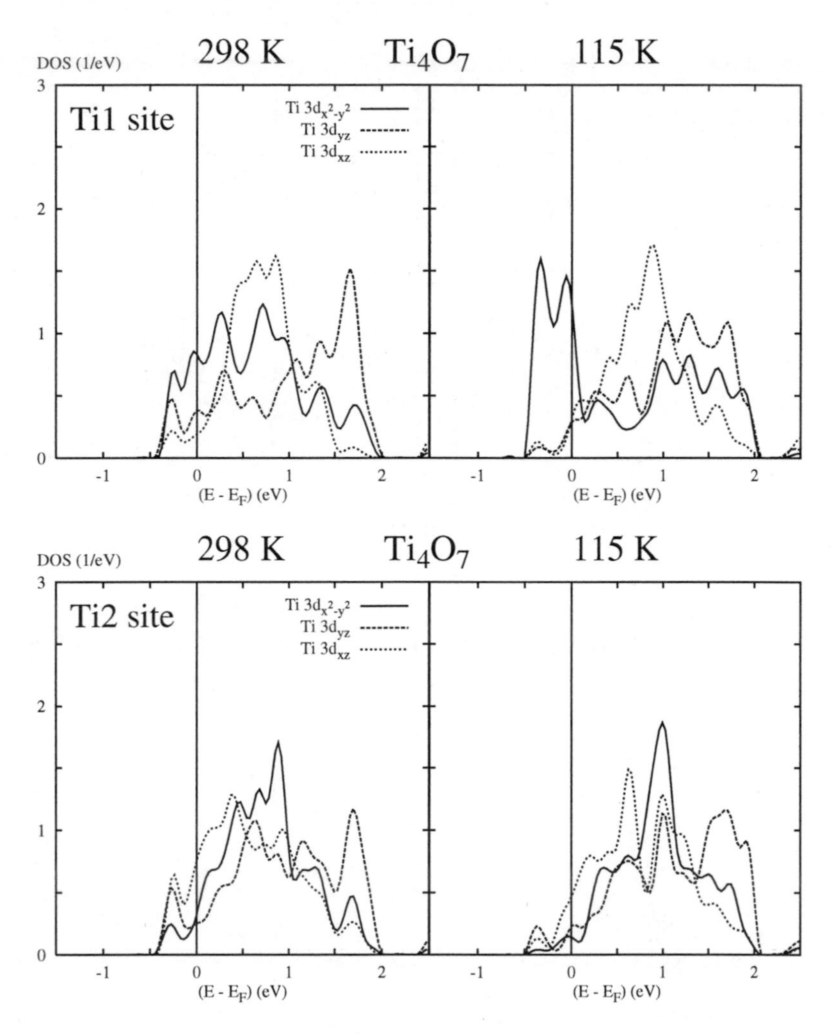

Figure 5.6: *Site-projected partial Ti 3d t_{2g} densities of states (DOS) per metal atom for the high as well as low temperature crystal structure of Ti_4O_7: sites Ti1 and Ti2. The crystal structure is similar to that of V_4O_7. The orbitals refer to the local rotated reference frame.*

of covalent Ti-O bonding.
Beyond the Ti-Ti dimerization, zigzag-type in-plane displacements of the titanium sites resemble the V_4O_7 findings in almost all respects. The distances between titanium atoms and octahedral centers amount to 0.31 Å/0.33 Å (Ti1), 0.33 Å/0.40 Å (Ti2), 0.16 Å/0.18 Å

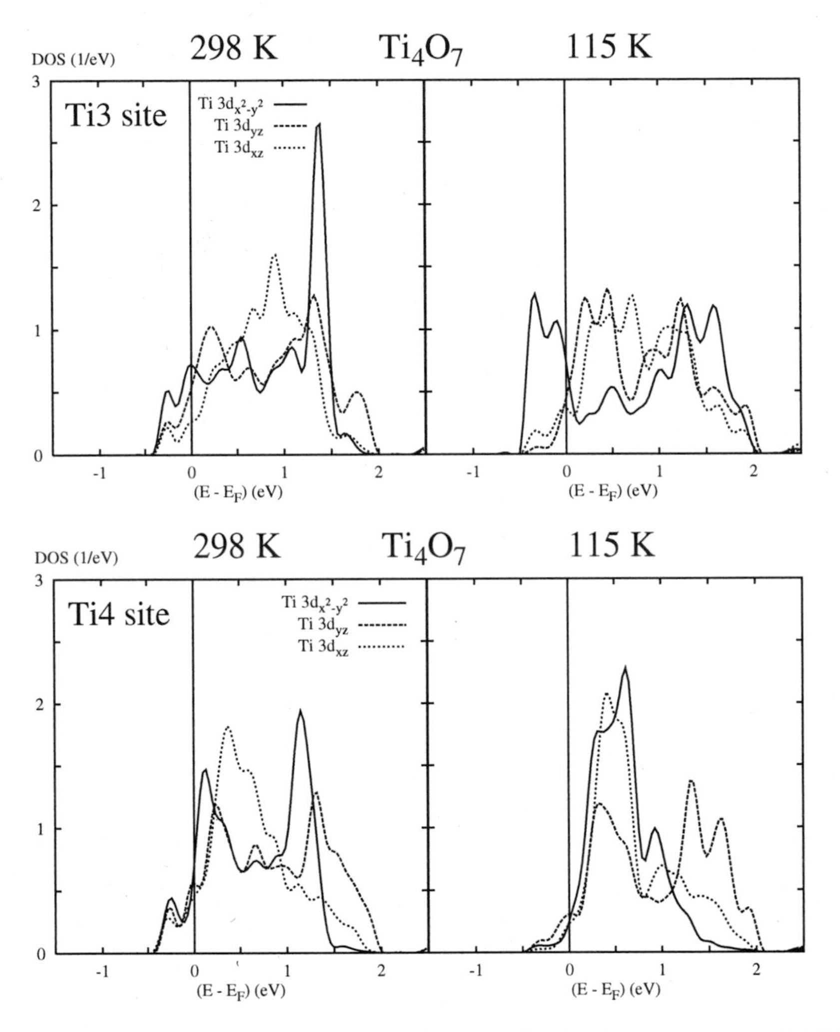

Figure 5.7: *Same representation as used in figure 5.6, but for the titanium sites Ti3/Ti4.*

(Ti3), and 0.09 Å/0.21 Å (Ti4) in the high/low temperature structure. Compared to V_4O_7 the Ti2 and Ti4 atoms reveal an increased distance in the insulating phase. A small rotation of the titanium 4-chains away from the c_{prut}-axis gives rise to an increased lateral displacement of the chain end atoms Ti1 and Ti2. Via the rotation the mutual distance of the latter sites grows along a_{prut}, which reflects the tendency of the titanium atoms in Ti_2O_3 to form anti-pairs along the c_{hex}-direction of the corundum structure. As expected,

the major contribution to the zigzag-type displacements in both Ti_4O_7 configurations is oriented perpendicular to c_{prut}. By virtue of the correspondency V1↔Ti2, V2↔Ti1 it is not surprising to find the shifts of the Ti1 atoms parallel to the apical axis of the local octahedron, i.e. parallel to the local z-axis. The Ti2 sites move perpendicular to both c_{prut} and their z-axis. Hence even the details of the atomic coordinations of V_4O_7 and Ti_4O_7 are most similar. Only one striking difference appears: in the low temperature configuration the strength of the in-chain Ti4-Ti4 overlap is strongly reduced. This fact is not only the consequence of an increased bond length but also traces back to an antiparallel shift of the Ti4 atoms along the local apical axis; thus a kind of kink evolves at the center of the chain. To a much lesser degree a similar oriented kink is inherent in the 1-3 titanium chain of the high temperature phase and disappears at the transition.
Because of σ-type metal-metal overlap along the c_{prut}-axis the chain center atoms display distinct splitting of the $d_{x^2-y^2}$ DOS in the high temperature phase, see figure 5.7. In the case of site Ti4 the double peak structure with the bonding peak at roughly 0.1 eV and the antibonding peak near 1.2 eV is clearly visible. For Ti3 a sharp antibonding peak is located at 1.4 eV, whereas the bonding contributions are rather broad and extend from the Fermi level up to approximately 0.6 eV. The deviations from a double peak structure may be a consequence of the modified in-chain titanium overlap due to the kink in the 1-3 chain. As expected from the previous structural considerations the chain end atoms (figure 5.6) are less affected by in-chain Ti-Ti bonding and hence reveal a compact $d_{x^2-y^2}$ DOS. At low temperatures the behaviour of both titanium chains changes substantially. The increased Ti1-Ti3 dimerization leads to an enhanced bonding-antibonding splitting of the Ti1 and Ti3 $d_{x^2-y^2}$ densities of states, which principally resemble one another. Due to the substantially reduced Ti-Ti overlap in the 2-4 chains the $d_{x^2-y^2}$ DOS of both Ti2 and Ti4 sharpens considerably at the MIT. While the 1-3 titanium chain clearly mirrors the results for the 2-4 vanadium chain of V_4O_7, the localization of all four $d_{x^2-y^2}$ orbitals in the 2-4 titanium chain is a new feature.
Small changes at the phase transition affect the titanium d_{xz} partial DOS. Minor upshifts of spectral weight arise from growing titanium displacements perpendicular to c_{prut}. Moving metal atoms laterally off the centers of gravity of their surrounding oxygen octahedra increases the overlap between Ti $3d$ and O $2p$ orbitals and thus the bonding-antibonding splitting. As is usual for the vanadium/titanium Magnéli compounds the structural transformation leaves the oxygen sublattice almost unchanged. Hence we can interpret the d_{xz} densities of states in terms of the zigzag-type distortions affecting the titanium chains. In the case of atom Ti3 such effects are barely visible due to a reduced shift. For site Ti4 a sharpening of the d_{xz} DOS results from the decreasing in-chain Ti-Ti overlap. In general, metal-metal bonding is less important for d_{xz} orbitals due to their π-type overlap. Thus splitting in bonding and antibonding branches is not significant.
For the titanium d_{yz} orbitals the analysis of the DOS curves is a greater challenge since different types of hybridization are involved. In principle, these orbitals are affected by metal-metal overlap across octahedral edges both along a_{prut} and b_{prut}. The zigzag-type metal displacements cause energetical upshift of the d_{yz} states. Consequently, the DOS directly below the Fermi level is reduced in the low temperature phase. Sites Ti3 and Ti4 do not mediate Ti-Ti interaction along a_{prut} since they lack nearest titanium neighbours in this direction. Hence the pronounced double peak structure of both the Ti3 and the

Ti4 d_{yz} DOS implies bonding-antibonding splitting due to Ti1-Ti3 and Ti2-Ti4 overlap, respectively, which is effective at high and low temperatures. Transferring our knowledge of the atomic arrangements in V_4O_7 to the titanium system allows us to identify Ti1-Ti3 as well as Ti2-Ti4 pairs inducing bonding across shared octahedral faces parallel to b_{prut}. In contrast to the central atoms, the chain end atoms Ti1 and Ti2 take part not only in Ti1-Ti3/Ti2-Ti4 overlap parallel to b_{prut} but also in Ti1-Ti2 bonding along a_{prut}. Therefore the d_{yz} partial DOS reflects features reminiscent of the corresponding Ti2/Ti3 DOS or Ti1/Ti4 DOS. The triangular shape of the d_{yz} DOS curves mirrors the characteristics known from V_4O_7. Due to their increased lateral displacements the chain end atoms are affected by stronger metal-oxygen interaction below the MIT. Thus spectral weight shifts to higher energies, which is observed in figure 5.6.

Summarizing, the electronic properties of Ti_4O_7 are strongly influenced by the local environments of the atomic sites. Overlap of Ti $3d$ and O $2p$ orbitals locates the Ti $3d$ t_{2g} group of bands around the Fermi energy, whereas the details of the electronic structure trace back to the metal-metal coordination. We observe remarkable bonding-antibonding splitting of the chain center $d_{x^2-y^2}$ states. At low temperatures the splitting even increases at the Ti3 site but vanishes at the Ti4 site. Due to strong in-chain Ti1-Ti3 bonding the low temperature Ti1 $d_{x^2-y^2}$ DOS splits up and resembles the shape of the Ti3 DOS. Furthermore, the Ti2 $d_{x^2-y^2}$ DOS reveals a compact structure, which even sharpens below the transition since the orbital overlap to adjacent titanium sites decreases. For the same reason the low temperature Ti4 $d_{x^2-y^2}$ DOS is dominated by a single sharp peak. By virtue of their localized nature the $d_{x^2-y^2}$ orbitals of the 2-4 chain are susceptible to electronic correlations. Both d_{xz} and d_{yz} densities of states respond to lateral displacements of the titanium atoms. Moving them off their octahedral centers enforces an energetical upshift of the e_g^π states due to increased Ti-O overlap. Each site-projected d_{yz} DOS is influenced by Ti-Ti interaction across octahedral faces. The two peak structures of the chain center d_{yz} densities of states reflect bonding parallel to b_{prut}. Ti1 and Ti2 participate also in d_{yz} overlap along a_{prut} and hence resemble the chain end sites of V_4O_7.

The increase and decrease of the occupied Ti1/Ti3 and Ti2/Ti4 $3d$ t_{2g} DOS traces back to charge ordering induced by modified Ti-O bond lengths in the low temperature crystal structure. In the 2-4 chains smaller Ti-O distances lead to stronger bonding-antibonding splitting of the hybridized Ti $3d$ and O $2p$ states and hence upshift of the antibonding bands. This effect is supported by displacements of the titanium atoms perpendicular to the metal chains. Eventually, we have almost completely depopulated Ti1/Ti3 $3d$ states and d^0 titanium configurations. In contrast, on the 1-3 chains longer Ti-O distances lead to reduced bonding-antibonding splitting, which results in energetical lowering of the Ti $3d$ bands and occupation close to d^1. Corresponding changes in the band structure are found in figure 5.3, especially near the Fermi level. The low temperature band structure is characterized by two split-off bands at the lower edge of the t_{2g} group. As expected from the DOS, the analysis of the wave functions reveals an almost pure Ti1/Ti3 $d_{x^2-y^2}$ character of these weakly dispersing bands. Additional bands of the same symmetry, showing also reduced dispersion, are found at the upper edge of the t_{2g} group. This supports our previous discussion in terms of bonding-antibonding splitting.

We now analyze the calculated LDA valence charges given in table 5.2. Recall the limitations of assigning charge to specific atomic sites. While table 5.2 gives rather similar

valence charges for all the titanium sites in the high temperature structure, charge ordering is found below the phase transition. Compared to the 2-4 titanium chain we observe roughly 0.3 additional electrons for each atom of the 1-3 chain, which corresponds to a charge difference of 0.6 electrons per formula unit. Importantly, due to modifications of the Ti-O overlap at the transition the hybridization between the Ti $3d$ and O $2p$ states is different for high and low temperature Ti_4O_7. Consequently, it is disadvantageous to calculate the charge transfer by means of the Ti $3d$ valence charges from table 5.2 since they result from integrating the full Ti $3d$ DOS up to the Fermi level. Modified titanium contributions to the oxygen dominated energy range would falsify the charge transfer results. In order to compare the local charges of the high and the low temperature phase we therefore consider only the t_{2g} region in the integration. In doing so we find a charge difference of about 0.9 electrons per formula unit, which is relevantly more than the 0.6 electrons calculated from the valence charges. Of course, this argumentation applies also to the discussion of the vanadium Magnéli phases in the preceding chapter. However, for the vanadium systems charge transfer effects are smaller. The calculated charge transfer in low temperature Ti_4O_7 qualitatively agrees with the results of M. Marezio *et al.* [88], who used a simple ionic picture and argued for 3+ and 4+ valency stages at the Ti1/Ti3 and Ti2/Ti4 sites, respectively. Although this pattern is confirmed by the LDA calculation the resulting charge difference is much smaller.
In general, the local electronic properties of Ti_4O_7 resemble those of the rutile-type compounds TiO_2 and VO_2, where the metal atoms have d^0 and d^1 configuration, respectively. The titanate is insulating because of its empty d shell and preserves the rutile structure down to lowest temperatures, maintaining especially the equidistant spacing of the metal atoms. In contrast, vanadium dioxide is characterized by its embedded Peierls instability discussed at length in chapter 3. In particular, the vanadium atoms are subject to pairing parallel to c_{prut} and zigzag-type antiferroelectric displacements perpendicular to this axis, which yields splitting of the $d_{x^2-y^2}$ bands and energetical upshift of the d_{xz} and d_{yz} states. One may therefore regard low temperature Ti_4O_7 to be composed of TiO_2 and VO_2-like chains showing the same behaviour as the respective dioxide. Common to VO_2 and Ti_4O_7 is a small band overlap at the Fermi energy present in the low temperature results, which we attribute to the LDA. However, this failure does not undermine an understanding of the MITs, which arise from orbital order and bonding-antibonding splitting of the $d_{x^2-y^2}$ bands because of structural modifications and strong electron-phonon coupling. Ti_4O_7 is characterized by distinct charge ordering leading to separated d^0 and d^1 titanium chains. Thus charge ordering forms the basis for the efficacy of the Peierls-like mechanism.
In conclusion, the phase transition traces back to a complex interplay of different types of ordering phenomena. While the titanium sites reveal similar $3d$ t_{2g} occupations at high temperatures, the low temperature phase is characterized by metal chains comprising either Ti d^0 or Ti d^1 configurations. Charge ordering is accompanied by orbital ordering in the d^1 chains and bonding-antibonding splitting of the respective $d_{x^2-y^2}$ states coming with Ti-Ti dimerization and zigzag-type antiferroelectric displacements. While the distortions of the 1-3 chains are interpreted in analogy to VO_2, the 2-4 chains are subject to reduced in-chain metal-metal overlap at low temperatures. Thus the $d_{x^2-y^2}$ states localize and become susceptible to electronic correlations. All in all, the MIT of Ti_4O_7 arises from the instability of the d^1 titanium chains towards a Peierls-like distortion.

5.2 On the 1D bands of Ti_2O_3 and V_2O_3

Ti_2O_3 crystallizes in the corundum structure, which we have already discussed in the context of the isostructural compound V_2O_3. Between 400 K and 600 K titanium sesquioxide undergoes a gradual MIT without an accompanying symmetry breaking lattice distortion. In contrast to V_2O_3 its low temperature insulating state is observed to be non-magnetic. The conventional explanation of this fact is based on a splitting of the t_{2g} states. In the corundum structure the t_{2g} manifold arising from crystal field effects due to the oxygen octahedra splits up into bonding and antibonding a_{1g} and e_g^π levels. The a_{1g} orbitals have $d_{3z^2-r^2}$ character and form strong bonds between pairs of titanium atoms in face-sharing octahedra parallel to c_{hex}, which results in bonding a_{1g} and antibonding a_{1g}^* bands bracketing the e_g^π and $e_g^{\pi *}$ bands. An insulating energy gap is expected to occur between the occupied a_{1g} and unoccupied e_g^π subbands. According to this picture, which traces back to L. L. Van Zandt, J. M. Honig, and J. B. Goodenough [123], the increase of the ratio c/a of the corundum lattice constants with increasing temperature reduces the bonding-antibonding splitting of the a_{1g} bands and hence promotes the collapse of the insulating gap. Experimental findings confirm the thermal closure of the band gap. For details see the review by M. Imada *et al.* [34] and the references therein.
Little is known about the high temperature metallic phase, which is assumed to display strongly correlated electron behaviour. On the basis of LDA band structure calculations using the linear augmented plane wave method L. F. Mattheiss [124] proposed the importance of electronic correlations as a possible source of the MIT. His calculation results in a partially filled t_{2g} complex near the Fermi level originating from overlapping a_{1g} and e_g^π subbands. Decreasing the c/a ratio beyond the observed low temperature value reduces but does not eliminate the a_{1g}-e_g^π overlap, in contrast to the suggestion of the previously introduced model. To open an insulating energy gap in this system an unphysically small Ti-Ti distance of 2.2 Å parallel to c_{hex} is required, which precludes a simple band explanation of the MIT. In order to estimate the influence of structural changes on the phase transition, W.-D. Yang [56] artificially exaggerated the modifications. Applying a Ti-Ti distance of 2.39 Å, intead of the experimental 2.59 Å, yields a band gap. For the actual low temperature structure the a_{1g} and a_{1g}^* states expectedly bracket the e_g^π bands – but a finite DOS remains at the Fermi energy. The result changes for the hypothetical structure, where the a_{1g}, e_g^π and a_{1g}^* groups of bands clearly separate. Therefore not only the metal-metal dimerization but also the rest of the structural transformation is important for the MIT of Ti_2O_3, particularly the Ti-O bond lengths. The lacking optical band gap in the case of the actual structure is assigned to LDA shortcomings. Special importance of the Ti-O bonding follows from the investigation of the total energy. While varying the Ti-Ti distances parallel to c_{hex} marginally modifies the stability of the crystal, the major part of the energy gain at the phase transition traces back to an instability in the geometry of the TiO_6 octahedra. Since LDA+U calculations allow for an opening of the optical band gap the MIT of Ti_2O_3 might be interpreted as a result of structural modifications and electronic interactions. Recent cluster LDA+DMFT calculations assuming moderate Coulomb interactions among the a_{1g} orbitals reproduced the insulating state [125].
The crystal structures of Ti_2O_3 at temperatures ranging from 296 K to 868 K have been determined by C. E. Rice and W. R. Robinson [126]. In order to account for the temper-

ature range of the MIT we subsequently use the data at 296 K and 621 K. The corundum structure of Ti_2O_3 is based on a trigonal lattice with space group $R\bar{3}c$ (D^6_{3d}). Often a non-primitive hexagonal cell containing six formula units is applied in the literature, whereas the primitive trigonal cell comprises two formula units. The hexagonal lattice constants at 296 K amount to $a_H = 5.1580$ Å and $c_H = 13.611$ Å. We find the vanadium atoms at the Wyckoff positions (12c): $\pm(0,0,z_{Ti})$, $\pm(0,0,1/2+z_{Ti})$, $\pm(-1/3,1/3,1/3)\pm(0,0,z_{Ti})$, and $\pm(-1/3,1/3,1/3)\pm(0,0,1/2+z_{Ti})$. Furthermore, the oxygen atoms occupy the positions (18e): $\pm(x_O,0,1/4)$, $\pm(0,x_O,1/4)$, $\pm(-x_O,-x_O,1/4)$, $\pm(-1/3,1/3,1/3)\pm(x_O,0,1/4)$, $\pm(-1/3,1/3,1/3)\pm(0,x_O,1/4)$, and $\pm(-1/3,1/3,1/3)\pm(-x_O,-x_O,1/4)$. The positional parameters are given by $z_{Ti} = 0.34469$ as well as $x_O = 0.31315$ and the atomic coordinates in the hexagonal notation refer to the primitive translations defined in equation (3.4). Following equations (3.5) to (3.8) we set up the trigonal unit cell as in the case of V_2O_3 and transform atomic positions given in terms of the hexagonal primitive translations into the trigonal representation. Accordingly, the titanium atoms in the trigonal lattice are located at the Wyckoff positions (4c): $\pm(z^*_{Ti},z^*_{Ti},z^*_{Ti})$ and $\pm(1/2+z^*_{Ti},1/2+z^*_{Ti},1/2+z^*_{Ti})$. Moreover, the oxygen atoms take the positions (6e): $\pm(x^*_O,1/2-x^*_O,1/4)$, $\pm(1/4,x^*_O,1/2-x^*_O)$, and $\pm(1/2-x^*_O,1/4,x^*_O)$. To complete the representation we denote the positional parameters $z^*_{Ti} = z_{Ti} = 0.34469$, $x^*_O = x_O + 1/4 = 0.56315$ as well as the lattice constants $a_T = 2.97797$Å, $c_T = 4.53700$ Å at low temperatures. In the high temperature structure one finds $a_H = 5.1260$ Å, $c_H = 13.878$ Å, $z_{Ti} = 0.34674$, $x_O = 0.31070$, $a_T = 2.95950$Å, $c_T = 4.62599$Å, $z^*_{Ti} = 0.34674$, and $x^*_O = 0.56070$.

Because Ti_2O_3 is isostructural to V_2O_3 the titanium atoms are octahedrally coordinated by six oxygen atoms. Moreover, the oxygen network is qualitatively equivalent to that of the vanadium Magnéli phases, including the end members VO_2 and V_2O_3. In particular, parallel to c_{hex} (a_{prut}) the oxygen octahedra are linked via faces. Every third octahedron stays empty, whereas the others are filled with titanium atoms. Perpendicular to c_{hex} the titanium atoms form hexagonal structures, allowing for a description of Ti_2O_3 in terms of a non-primitive hexagonal unit cell. The schematical representation of the V_2O_3 structure shown in figure 3.10 applies also to titanium sesquioxide. Due to the trigonal symmetry each titanium atom exhibits three identical in-plane Ti-Ti distances of 2.98 Å (621 K) or 2.99 Å (296 K). Moreover, there is one shorter bond of 2.69 Å (621 K) or 2.58 Å (296 K) to the adjacent titanium site along c_{hex}. In contrast to the PM-AFI transition of vanadium sesquioxide the lattice symmetry of Ti_2O_3 is not changed at the phase transition. In order to fix the pseudorutile reference frame in the case of titanium sesquioxide we choose one of the three equivalent in-plane Ti-Ti bonds as the c_{prut}-axis.

From simple electrostatic considerations it is not surprising that the hexagonal (in-plane) titanium structures deviate from being planar. Due to the strong and rather unscreened ionic Ti-Ti interaction through octahedral faces along c_{hex} the metal pairs are subject to an anti-dimerization, as in the case of V_2O_3. The alternating Ti-Ti distances in the c_{hex}-direction thus do not fulfill a ratio of 2:1. Instead of the expected in-pair distances 2.31 Å and 2.27 Å we find the values 2.69 Å and 2.58 Å above and below the MIT, respectively. Compared to vanadium sesquioxide the magnitude of the metal-metal anti-dimerization is about the same in Ti_2O_3, but the changes at the phase transition behave contrarily. While the c_{hex} V-V distance in V_2O_3 increases from 2.70 Å to 2.75 Å on going from the metallic to the insulating phase, we observe a significant reduction of the Ti-Ti anti-dimerization

in the low temperature phase of Ti_2O_3. Since the oxygen sublattices display only minor changes at the transition, the anti-dimerizations both in the vanadium and the titanium compound correspond to shifts of the metal atoms away from the octahedral centers. Except for small deviations in the case of monoclinic V_2O_3 the shifts are oriented along the c_{hex}-axis. More specifically, they amount to 0.18 Å and 0.21 Å for metallic and insulating vanadium sesquioxide as well as to 0.19 Å and 0.15 Å in the titanium compound. For high temperature V_2O_3 we calculate 2.90 electrons of V $3d$ valence charge in an atomic sphere with radius 2.34 a_B and for the low temperature structure 2.87 electrons in a sphere with radius 2.35 a_B. For Ti_2O_3 the Ti $3d$ charge amounts to 1.90 electrons (at 621 K) or 1.91 electrons (at 296 K), where the numbers refer to atomic spheres with radii 2.44 a_B. Thus we observe little difference between the metallic and insulating phases of V_2O_3 and Ti_2O_3 from ionic considerations. In both compounds the changes of the oxygen sublattice at the transition are rather small. Hence we may relate the modification of the anti-dimerization to changes in the covalent metal-metal bonding. While covalent bonding seems to weaken in the low temperature phase of V_2O_3, it strengthens for Ti_2O_3.
The comparison of the d^1 system Ti_2O_3 with the isostructural d^2 compound V_2O_3 yields important conclusions about the structural origin of the electronic properties. In particular, it allows for further insight into the shapes of the t_{2g} DOS in the vicinity of the Fermi level. To this end we now discuss the LDA results for Ti_2O_3 analogous with our investigation of the vanadium compound, see section 3.4. Figure 5.8 displays the electronic bands arising from the crystallographic data of the corundum structure at 621 K and 296 K. As in the case of corundum V_2O_3 we depict the bands along selected high symmetry lines in the first Brillouin zone of the trigonal (rhombohedral) lattice, see figure 3.12.
The structural similarity between Ti_2O_3 and V_2O_3 implies at least the accordance of the gross features of their band structures. This is confirmed since we observe the same three groups of bands, which appear for the vanadium system. In the case of the low temperature structure they occupy the energy intervals from -9.0 eV to -4.5 eV, from -0.8 eV to 2.0 eV, and from 2.3 eV to 4.0 eV. Only minor changes of the energetical positions are found at low temperatures, except for a widening of the t_{2g} group of bands, which reaches from -0.9 eV to 2.3 eV. Note the titanium $4s$ states at energies higher than 3.9 eV. All these findings agree well with the results of L. F. Mattheiss [124]. In comparison to V_2O_3 the separation between the energetically lowest and the middle group of bands has grown by about 0.7 eV, which indicates a stronger bonding-antibonding splitting and hence an inreased overlap between Ti $3d$ and O $2p$ states. The energetical upshift of the t_{2g} group is due to a reduced electron count in the d^1 system Ti_2O_3. As known from corundum V_2O_3 the unit cell of the titanium analog contains two formula units. Thus we have 18, 12, and 8 bands in the lowest, middle, and highest group of bands in figure 5.8. Concluding, the band structure findings allow for an interpretation of titanium and vanadium sesquioxide in terms of a similar molecular orbital picture.
In the partial Ti $3d$ and O $2p$ densities of states displayed in the upper part of figure 5.9 strong similarities to vanadium sesquioxide are visible. One actually finds three distinct structures in the DOS reflecting the three groups of states mentioned above. While the lowest group primarily traces back to O $2p$ orbitals, the groups at and above the Fermi energy are mainly due to Ti $3d$ states. Hybridization between oxygen and titanium yields Ti $3d$ and O $2p$ admixtures in the energy range dominated by the respective other states.

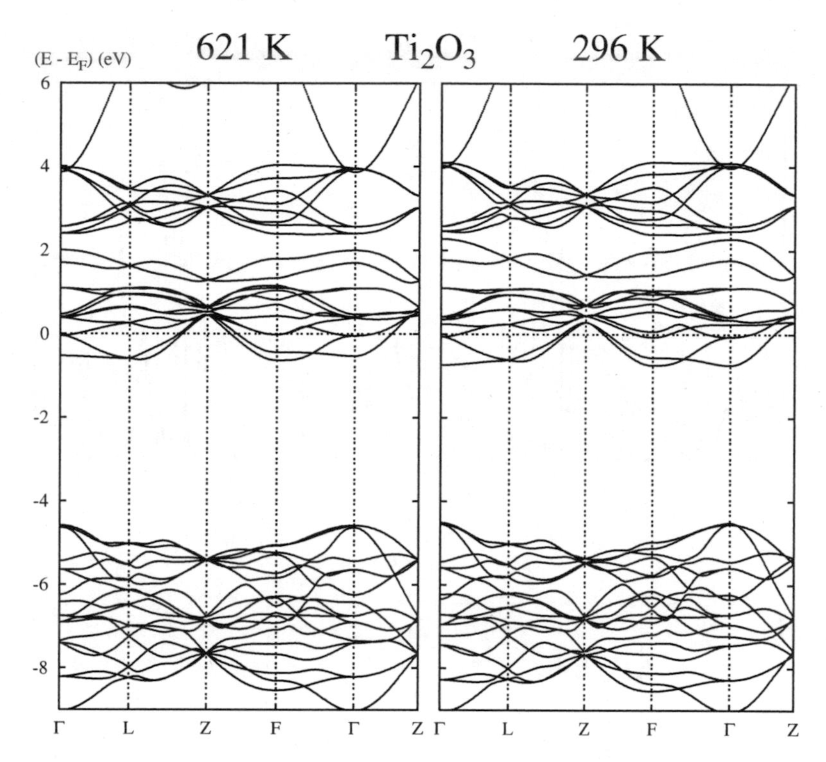

Figure 5.8: *Electronic bands of Ti_2O_3 both at 621 K and at 296 K displayed along selected symmetry lines in the first Brillouin zone of the trigonal lattice as depicted in figure 3.12.*

As expected, the mixing is stronger in the σ and σ^*-regions. Concerning the magnitude of the p-d hybridization Ti_2O_3 and V_2O_3 behave almost equivalent.
The partial DOS of the t_{2g} group is shown in the lower part of figure 5.9 – separated into its symmetry components. In addition, there are small t_{2g} contributions in the e_g^σ energy range above 2.3 eV. As in the case of vanadium sesquioxide they mainly trace back to the titanium anti-dimerization along $c_{\rm hex}$. Due to shifts off the ocahedral centers the coordination of titanium in respect to oxygen is perturbed and thus the separation in t_{2g} and e_g^σ states is no longer ideal. Only minor differences are visible in figure 5.9 distinguishing the DOS curves of the three symmetry components, which applies to both crystal structures considered. Similar to V_2O_3 the electronic structure of Ti_2O_3 barely responds to modifications of the crystal structure accompanying the MIT. The similarity of the t_{2g} DOS curves is a consequence of the trigonal symmetry of the corundum structure. Obviously, there is no indication of an energy gap in the calculation for the low temperature phase. We again attribute this fact to possible influence of electronic correlations, which are not fully accounted for by the LDA. The representation used for the data in figure 5.9 refers

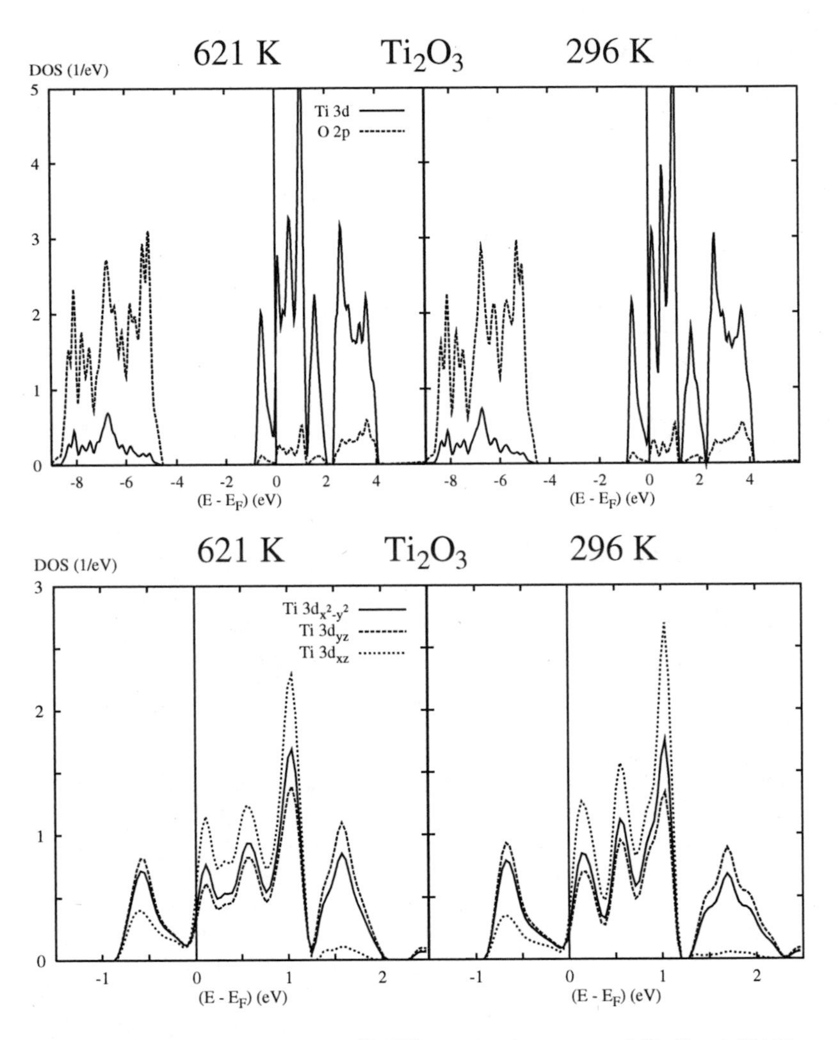

Figure 5.9: *Partial densities of states (DOS) per titanium atom of Ti_2O_3 at 621 K as well as 296 K. The first row presents a comparison of the partial Ti 3d and O 2p DOS resulting from the high and the low temperature crystal structure. In the second row the partial Ti 3d t_{2g} DOS is shown in detail. Here the orbitals refer to the local rotated reference frame.*

to the rotated reference, but due to the trigonal symmetry of the corundum structure an alternative description may be useful. As the molecular orbital picture predicts splitting of the Ti 3*d* t_{2g} states into a_{1g} and e_g^{π} contributions we decompose the DOS analogously,

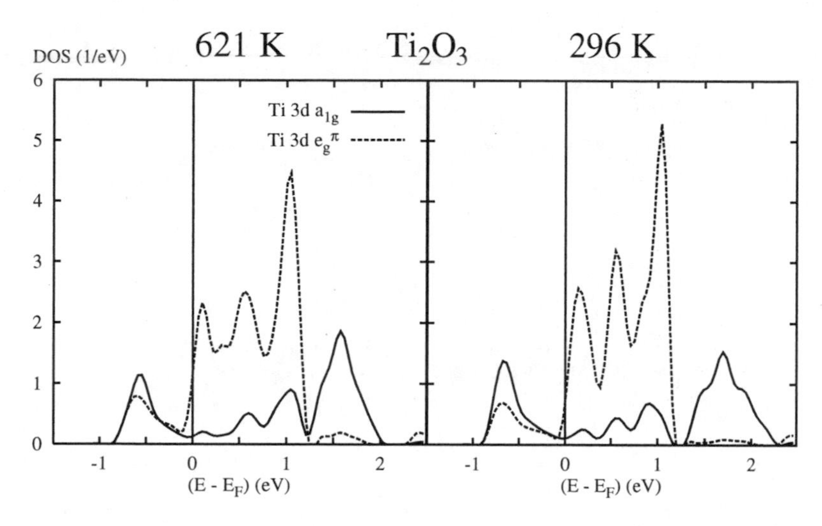

Figure 5.10: *Partial Ti 3d a_{1g} and Ti 3d e_g^π densities of states (DOS) per titanium atom of Ti_2O_3 both at 621 K and 296 K; alternative representation of the findings from figure 5.9.*

see figure 5.10, which allows us to indicate one-dimensional a_{1g} states. By comparing the results to the expectation from the molecular orbital picture we identify the two distinct peaks at about -0.6 eV and 1.6 eV (621 K) as the bonding and antibonding branches due to the interaction of nearest vanadium neighbours along c_{hex}. Because the titanium antipairing is reduced at low temperatures, the bonding-antibonding splitting grows and the DOS peaks shift to -0.7 eV and 1.7 eV. Confirming our expectation the bonding peak is found below the Fermi energy, whereas the antibonding states remain empty. Consistent with the unequally distributed weights of the bonding and the antibonding peak in V_2O_3, figure 5.10 reveals at both temperatures more spectral weight in the 1.6 eV peak than in the 0.6 eV peak, which we will comment on later. Induced by the diminished weight of the bonding a_{1g} branch, the 3*d* valence charge occupies not only a_{1g} but also some e_g^π states. Around 1.0 eV both DOS curves in figure 5.10 reveal a strong e_g^π peak combined with a smaller a_{1g} peak, which traces back to hybridization between a_{1g} and e_g^π states. The same applies to the a_{1g} states in the interstitial energy region from -0.3 eV to 1.2 eV.
We investigate the origin of the Ti_2O_3 a_{1g} DOS at 296 K by means of the weighted band structure given in figure 5.11. As the following conclusions are also true for the high temperature phase we omit to display the 621 K weighted band structure. As a consequence of strong structural similarities to V_2O_3 we obtain a close relation of the depicted band structure to that given in figure 3.15. The length of the bars attached to the bands represents the magnitude of the a_{1g} contributions to the particular states. In contrast to the bands shown in figure 5.8 we do not use the first Brillouin zone of the primitive trigonal lattice in figure 5.11. Instead, these bands refer to the non-primitive hexagonal representation as defined by equation (3.4) and the first Brillouin zone of the hexagonal lattice as

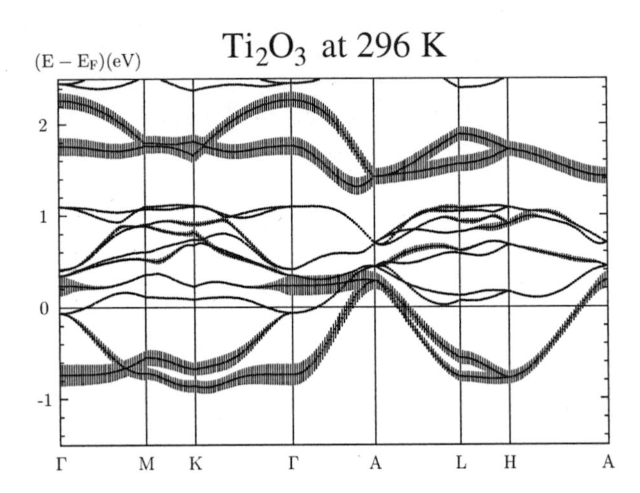

Figure 5.11: *Weighted electronic bands of Ti_2O_3 at 296 K shown along selected symmetry lines in the first Brillouin zone of the hexagonal lattice, compare figure 3.12. The width of the bars included for each band and k-point indicates the contribution due to the Ti 3d a_{1g} orbital of the atom located at site (z^*_{Ti},z^*_{Ti},z^*_{Ti}) relative to the local rotated reference frame.*

illustrated in figure 3.12. Due to the hexagonal-type arrangement of the titanium atoms in Ti_2O_3 the high symmetry lines of the hexagonal Brillouin zone reflect special directions in the titanium sublattice. In particular, the reciprocal k_z-axis is oriented parallel to the principal axes of the a_{1g} orbitals. Reflecting our results for the vanadium compound, the highlighted a_{1g} bands in figure 5.11 embrace the remaining e_g^π bands. The latter are represented by points without bars in the t_{2g} energy range. Similar to V_2O_3 the dispersion of the e_g^π bands is rather isotropic, whereas the strongest dispersion of the a_{1g} bands appears along the c_{hex}-direction (Γ-A). The antibonding bands at 1.8 eV and 2.3 eV (Γ point) have almost exclusively a_{1g} character throughout the Brillouin zone. They are completely separated from the e_g^π states, which does not apply to the bonding bands.
The aforementioned LDA calculation for hypothetical Ti_2O_3 by W.-D. Yang [56] reveals a complete separation of the bonding a_{1g} bands from the e_g^π group if the alterations of the insulating crystal structure are exaggerated. This observation confirms the importance of the particular Ti-O bond lengths for the electronic structure of Ti_2O_3. Not only the Ti-Ti distances but also the Ti-O interaction influences the separation of the bonding a_{1g} states. Modifying only the in-pair metal-metal bond length gives rise to a band gap only for very small Ti-Ti distances. While Ti-Ti dimerization along c_{hex} induces a splitting of the a_{1g} states into bonding and antibonding branches, Ti-O hybridization influences the shape of the bonding a_{1g} DOS. As the antibonding bands are clearly separated from the e_g^π states the effects here are much smaller. The latter is confirmed by small e_g^π contribution in the a_{1g} energy range above roughly 1.2 eV, see figure 5.10. Hence one may interpret the a_{1g} states in the energy interval below 1.2 eV, not only the ponounced peak at −0.6 eV, as the

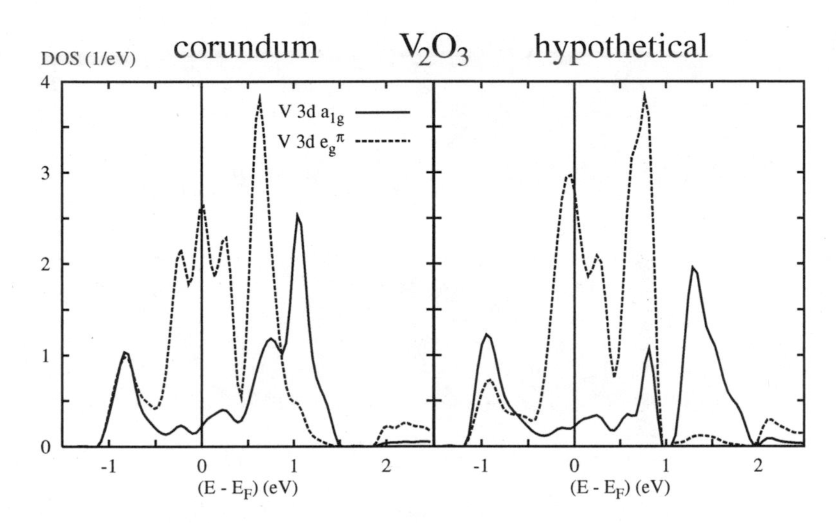

Figure 5.12: *Partial V 3d a_{1g} and V 3d e_g^π densities of states (DOS) per vanadium atom of corundum and hypothetical V_2O_3. In the case of the real crystal structure the V-V distance along c_{hex} amounts to 2.70Å, whereas it is reduced to 2.51Å in the hypothetical calculation.*

bonding branch of the a_{1g} DOS. By virtue of Ti-O interaction the effective band width of the bonding a_{1g} states increases considerably. This interpretation can solve the problem of unequal distributed spectral weights between bonding and antibonding branches. Due to strong a_{1g}-e_g^π coupling the 1D character of the a_{1g} states of Ti_2O_3 is perturbed.
Applying our insights into the shape of the a_{1g} DOS of the titanium compound to V_2O_3 enables us to understand the a_{1g} DOS of the vanadium system. Figure 5.12 depicts on the left hand side the known t_{2g} DOS of high temperature V_2O_3 and on the right hand side a corresponding DOS calculated for an artificial crystal structure. The hypothetical structure arises from the actual corundum structure by decreasing the V-V bond length along c_{hex} from 2.70 Å to 2.51 Å while maintaining the lattice symmetry. We neither modify the trigonal lattice constants $a_T = 2.85875$ Å and $c_T = 4.66767$ Å nor the positional parameter of the oxygen atoms $x_O^* = 0.56164$ – see the discussion of corundum V_2O_3 in section 3.3. Only the positional parameter of the vanadium sites changes from $z_V^* = 0.34630$ to $z_V^* = 0.33970$. In figure 5.13 a weighted band structure for the hypothetical modification of corundum V_2O_3 is shown. The length of the bars added to the bands is proportional to the a_{1g} contributions to the particular states. To allow for comparison with the weighted band structures of real V_2O_3 (figure 3.15) as well as Ti_2O_3 (figure 5.11), the bands in figure 5.13 refer to the non-primitive hexagonal representation of the lattice. The hexagonal Brillouin zone is presented in figure 3.12.
While the e_g^π DOS arising from the hypothetical V_2O_3 structure is closely related to the actual e_g^π DOS, we recognize the following modifications in the case of the a_{1g} DOS: the peak at about -0.8 eV shifts to lower energies. In addition, the dominating peak around

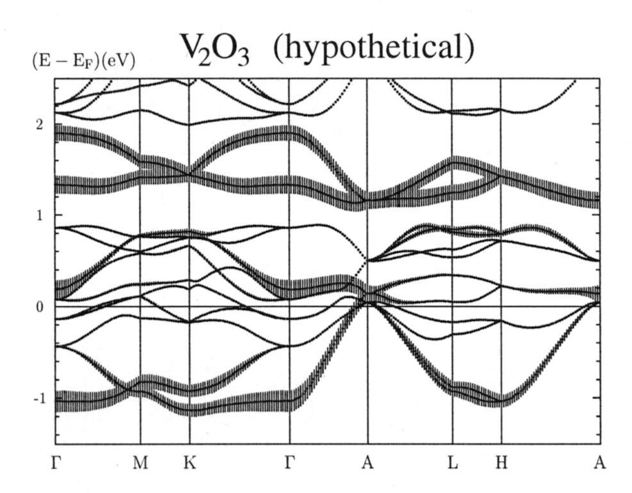

Figure 5.13: *Weighted electronic bands of hypothetical V_2O_3 given along selected symmetry lines in the first Brillouin zone of the hexagonal lattice, compare figure 3.12. The width of the bars included for each band and k-point indicates the contribution due to the V 3d a_{1g} orbital of the atom located at site (z^*_V,z^*_V,z^*_V) relative to the local rotated reference frame. Compared to real V_2O_3 the V-V distances along c_{hex} are downsized from 2.70 Å to 2.51 Å.*

1.0 eV splits up into two peaks located at 0.8 eV and 1.3 eV. Note that there is already a pronounced shoulder at 0.7 eV in the actual a_{1g} DOS. Moreover, the hypothetical V_2O_3 a_{1g} DOS resembles the findings for Ti_2O_3. Transferring our knowledge from the titanium to the vanadium compound implies regarding the hypothetical a_{1g} peak around 1.3 eV as antibonding branch induced by the V-V interaction along c_{hex}. The bonding branch comprises all the remaining a_{1g} states reaching from -1.2 eV to 1.0 eV. Due to the increased in-pair V-V interaction in the hypothetical structure the bonding-antibonding splitting is stronger, which enables comparison with the titanium system. The distinct a_{1g} peak at 0.8 eV for hypothetical V_2O_3 and the shoulder at 0.7 eV for real V_2O_3 trace back to the a_{1g}-e_g^π coupling. In each case a strong e_g^π peak exists at the same energy. For real V_2O_3 the wide bonding a_{1g} DOS even touches the antibonding peak thus making it impossible to understand the shape of the curve. The spectral weights of the entire bonding region and of the pronounced antibonding peak are almost equal. In the weighted band structure of hypothetical vanadium sesquioxide we clearly observe the split-off antibonding bands, whereas the bonding bands do not separate from the e_g^π states. Obviously, the weighted band structures of hypothetical V_2O_3 and real Ti_2O_3 agree amazingly well. In both cases the bonding branch of the a_{1g} states is strongly influenced by a_{1g}-e_g^π coupling.

Chapter 6

Octahedral Tilting in Perovskites

In this chapter we study the perovskite-like compounds $ACu_3Ru_4O_{12}$ (A=Na, Ca, Sr, La, Nd) by means of electronic structure calculations based on the density functional theory. Again we apply the augmented spherical wave method. We will find the electronic properties of $ACu_3Ru_4O_{12}$ to be strongly influenced by covalent bonding between the transition metal d and the O $2p$ orbitals. This insight forms the basis of investigating the origin as well as the consequences of the characteristical tilting of the RuO_6 octahedra. Particularly the copper atoms and their attempt to optimize the Cu-O bond lengths will arise as the driving force of the octahedral tilting. Indeed, octahedral tilting can be understood as a universal mechanism applicable to a large variety of multinary compounds. Parts of the subsequent discussion have been published previously [127].

6.1 Structural Considerations

Despite its chemical simplicity the perovskite structure ABX_3 allows for numerous crystallographic variations comprising many compounds [128, 129]. Most of the metallic ions in the periodic table can be inserted into the structure. Even though oxides and fluorides account for the majority of compounds, chlorides, hydrides, oxynitrides, and sulfids were also synthesized. As the perovskite structure is able to accommodate such a large variety of ions the materials reveal a great diversity of physical properties. Motivated by exciting dielectric, magnetic, electrical, optical, as well as catalytic features, huge interest in the perovskites emerged. Particularly their technological applicability turned the compounds into one of the most studied groups in materials science.
The ideal perovskite structure ABX_3 belongs to the cubic space group $Pm\bar{3}m$ (O_h^1). To be specific, the A cation is surrounded by twelve X anions in a cuboctahedral environment. Furthermore, the B cation is octahedrally coordinated by six X ions, which are surrounded by two B cations and four A cations. Figure 6.1 displays a simple cubic unit cell giving rise to the ideal perovskite. The most common way of describing the perovskite structure is a three dimensional cubic network of corner sharing BX_6 octahedra. Then the A cation occupies the center of a cube determined by eight of these corner sharing octahedra. An alternative representation of the crystal structure is based on a cubic close-packed array

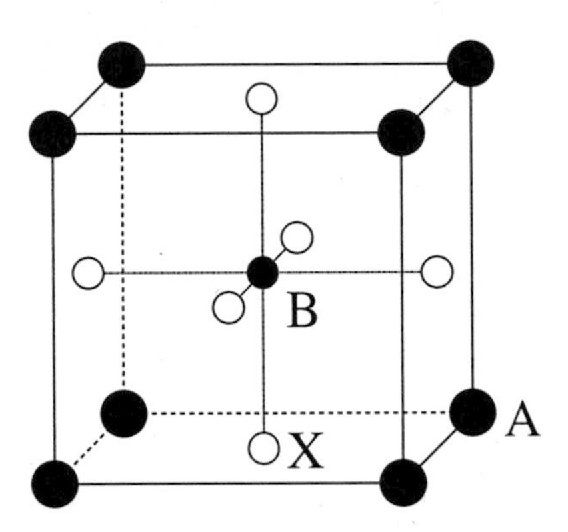

Figure 6.1: *Crystal structure of the (ideal) cubic perovskite – defined by the formula* ABX_3.

of X ions. Replacing every forth X ion with an A cation yields a close-packed AX_3 array. The arising octahedral holes must finally be filled with B atoms, if they do not border the A sites. Beyond a sizeable number of simple perovskites ABX_3 we observe an even larger amount of compounds with multiple ions substituted for one or more of the original types. These substitutions frequently occur on the cation sites. They change the symmetry and in many cases the size of the unit cell by means of a superstructure.
While the ideal perovskite structure is cubic, there are many materials revealing decisive deviations from this ideal configuration. In order to allow for an optimal material tailoring, many studies on the perovskite-related compounds concentrate on the interrelations between these deviations and the resulting physical properties. For instance, even small structural changes may have important effects on the electrical and magnetic behaviour. According to P. M. Woodward [128, 129] the crystallographic deviations are grouped into three mechanisms. Distortions of the characteristic BX_6 octahedra and cation displacements within the octahedra are mainly driven by electronic instabilities of the octahedrally coordinated ions. The Jahn–Teller distortion in $KCuF_3$ or the ferroelectric displacement of titanium in $BaTiO_3$ may serve as examples. Octahedral tiltings are the most common deviations and may be observed when the A cation is too small to fill the space between regularly ordered octahedra. In such situations, an octahedral tilting is the lowest energy distortion because it allows adjustment of the A-X distances, while leaving the first coordination sphere of the B cation intact. It changes the soft B-X-B bond angle rather than the B-X bond length. Interatomic forces driving octahedral tiltings are the subject of the following considerations.
In the context of octahedral tilting particularly the perovskite-related compounds of the kind $AA'_3B_4O_{12}$ are of interest. In this case the B-O-B bond angle is the relevant quantity for optimizing the A-O as well as the A′-O bond length. Studying these compounds promises deeper insights into the mechanisms driving the octahedral tilting because two bonds of different strength are optimized simultaneously. Furthermore, the material class is nowadays of major interest due to its extraordinary physical properties. For instance,

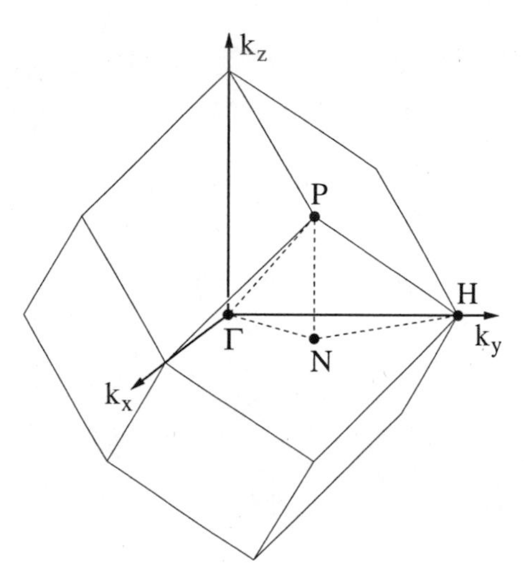

Figure 6.2: *First Brillouin zone of the body-centered cubic lattice. Here the included dashed lines mark the irreducible wedge of the zone and the labels denote points of high symmetry.*

$CaCu_3Mn_4O_{12}$ is a ferromagnetic semiconductor with high Curie temperature and large magnetoresistance [130]. $CaCu_3Ti_4O_{12}$ is special due to its unusually large low-frequency dielectric constant [131–133]. Because the dielectric constant in addition hardly changes between 100 K and 600 K the compound is most interesting for technological applications such as random access memories. Metallic ruthenates of the type $ACu_3Ru_4O_{12}$ (A=Na, Ca, La) are Pauli paramagnets [134]. Their metallic behaviour can be explained in terms of valence degeneracy between copper and ruthenium [133]. Due to Cu-Ru bonding both copper $3d$ and ruthenium $4d$ states are found close to the Fermi energy. Recently, crystal structures of the latter class of compounds were analyzed in detail [135], allowing for an investigation of the electronic properties and their interrelations to the structural arrangement by means of band structure calculations. Because the structural refinements pointed to considerable deviations between the measured atomic distances and those expected in the framework of a bond valence approach, this kind of study is desirable. The following analysis thus aims at an investigation of the relationship between the octahedral tiltings and the electronic features of the ruthenates $ACu_3Ru_4O_{12}$ (A=Na, Ca, Sr, La, Nd). As a result, we will be able to understand the specific bond lengths as well as the discrepancies between the measured interatomic distances and the values proposed by the bond valence approach in terms of the electronic properties and the type of chemical bonding realized in the material class.

The perovskite-related compounds $AA'_3B_4O_{12}$ crystallize in a body-centered cubic $2\times2\times2$ superstructure of the simple cubic perovskite structure. Figure 6.2 gives the first Brillouin zone of the body-centered cubic lattice. Because of the evolving superstructure the lattice symmetry in the class $AA'_3B_4O_{12}$ is reduced. Hence we have to deal with the space group

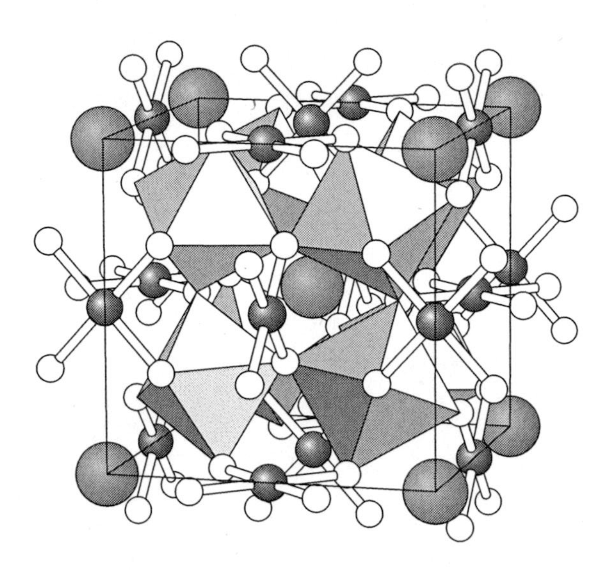

Figure 6.3: *Crystal structure of $ACu_3Ru_4O_{12}$ (A=Na, Ca, Sr, La, Nd). The body-centered cubic unit cell is a 2 × 2 × 2 superstructure of the simple cubic perovskite structure. Large and medium spheres denote A-sites and copper atoms, respectively. RuO_6 units are shown as octahedra with oxygen vertices (small spheres); ruthenium occupies the octahedral sites.*

$Im\bar{3}$ (T_h^5). Atoms of type A are located at the Wyckoff positions (2a): $(0,0,0)$, A′ atoms at the positions (6b): $(0,1/2,1/2)$, $(1/2,0,1/2)$, $(1/2,1/2,0)$, and B atoms at the positions (8c): $(1/4,1/4,1/4),(3/4,3/4,1/4),(3/4,1/4,3/4),(1/4,3/4,3/4)$. Oxygen is observed at the positions (24g): $\pm(x_O,y_O,0)$, $\pm(x_O,-y_O,0)$, $\pm(0,x_O,y_O)$ $\pm(0,x_O,-y_O)$, $\pm(y_O,0,x_0)$, $\pm(-y_O,0,x_0)$. The atomic arrangement is displayed in detail in figure 6.3. For the compounds $ACu_3Ru_4O_{12}$ the A sites are shown by large spheres, whereas medium spheres denote copper. The octahedra illustrate RuO_6 units with the oxygen atoms (small spheres) located at the vertices. Ruthenium atoms occupying the octahedral sites are omitted. A remarkable tilting of the oxygen octahedra is easily identified in figure 6.3. Importantly, this octahedral tilting is completely controlled by the oxygen positions – defined by only two parameters x_O and y_O. S. G. Ebbinghaus *et al.* [135] gave the values $x_O = 0.18080$, $y_O = 0.30862$ (Na), $x_O = 0.17320$, $y_O = 0.30348$ (Ca), $x_O = 0.17912$, $y_O = 0.30676$ (Sr), $x_O = 0.17618$, $y_O = 0.30541$ (La), and $x_O = 0.17174$, $y_O = 0.29841$ (Nd). Furthermore, for the cubic lattice parameter they found the values $a = 7.38489$ Å (Na), $a = 7.41871$ Å (Ca), $a = 7.44754$ Å (Sr), $a = 7.47800$ Å (La), and $a = 7.45780$ Å (Nd). All these numbers are very similar to those reported by M. Labeau *et al.* [134] as well as by M. A. Subramanian and A. W. Sleight [133]. Both the lattice constants and the positional parameters of the oxygen atoms reveal only small deviations for the listed compounds.
Obviously, the octahedral tiltings are similar for all the ruthenates under study. In good approximation the tilting is given by the parameters $x_O = 0.175$ and $y_O = 0.305$, which

	Na $3s^1$	Ca $4s^2$	Sr $5s^2$	La $5d^16s^2$	Nd $4f^46s^2$
Atom	Radius	Radius	Radius	Radius	Radius
A	2.8435	3.1125	3.2588	3.3660	3.2219
Cu	2.2314	2.2370	2.2562	2.2498	2.2619
Ru	2.4135	2.4385	2.4354	2.4188	2.4339
O	1.8362	1.8597	1.8574	1.8633	1.8562

	Na $3s^1$	Ca $4s^2$	Sr $5s^2$	La $5d^16s^2$	Nd $4f^46s^2$
Atom	Charge	Charge	Charge	Charge	Charge
A	0.1348	0.6268	0.7200	1.2029	4.5226
Cu	9.0277	9.0341	9.0212	9.0569	9.0702
Ru	5.4946	5.5958	5.6444	5.6221	5.6668
O	4.0225	4.0770	4.0786	4.0881	4.0768

Table 6.4: *Atomic sphere radii (in a_B) and calculated LDA valence charge (Na 3s, Ca 3d, Sr 4d, La 5d, Nd 4f+5d, Cu 3d, Ru 4d, O 2p) for $ACu_3Ru_4O_{12}$ (A=Na, Ca, Sr, La, Nd).*

almost satisfy the relation $x_O + y_O = 0.5$. In contrast, the ideal case without tilting corresponds to the numbers $x_O = y_O = 0.25$. From figure 6.3 it becomes clear that the realized deviations from these ideal parameters yield a rotation of the oxygen octahedra around the diagonals of the (cubic) unit cell. As a consequence, the cuboctahedrally coordinated sites of the ideal perovskite structure become separated into A and A′ sites. While the A sites comprise one quarter of the ideal perovskite positions, the A′ sites take three quarters. Due to an oxygen dislocation by means of an octahedral distortion the coordination geometry of the A sites changes from cuboctahedral to icosahedral. Simultaneously, the cuboctahedra of the A′ cations are subject to large distortions. Therefore three different A′-O distances of roughly 2.0 Å, 2.8 Å, and 3.2 Å appear. In the case of the ideal cubic perovskite all twelve metal-oxygen distances amount to about 2.6 Å. The four short A′-O bond lengths give rise to an approximately square-planar coordination of the A′ cations. This kind of local environment provides an ideal geometry for Jahn–Teller active ions as Cu^{2+}. Thus the typical B-type cation copper is located at the A′ site in $ACu_3Ru_4O_{12}$. In figure 6.3 all the CuO_4 squares are marked by thin sticks connecting copper and oxygen. The following LDA band structure calculations were performed in the framework of the density functional theory and the local density approximation using the augmented spherical wave method. To set up the unit cell for the calculation we start with the primitive translations of the simple cubic lattice

$$\mathbf{a}_C = \begin{pmatrix} a \\ 0 \\ 0 \end{pmatrix}, \quad \mathbf{b}_C = \begin{pmatrix} 0 \\ a \\ 0 \end{pmatrix}, \quad \mathbf{c}_C = \begin{pmatrix} 0 \\ 0 \\ a \end{pmatrix}. \tag{6.1}$$

By means of the transformation [66]

$$\begin{pmatrix} \mathbf{a}_{BC} \\ \mathbf{b}_{BC} \\ \mathbf{c}_{BC} \end{pmatrix} = \frac{1}{2} \begin{pmatrix} -1 & 1 & 1 \\ 1 & -1 & 1 \\ 1 & 1 & -1 \end{pmatrix} \begin{pmatrix} \mathbf{a}_C \\ \mathbf{b}_C \\ \mathbf{c}_C \end{pmatrix} \tag{6.2}$$

we gain the primitive translations of the body-centered cubic lattice in Carthesian coordinates

$$\mathbf{a}_{BC} = \frac{a}{2}\begin{pmatrix} -1 \\ 1 \\ 1 \end{pmatrix}, \quad \mathbf{b}_{BC} = \frac{a}{2}\begin{pmatrix} 1 \\ -1 \\ 1 \end{pmatrix}, \quad \mathbf{c}_{BC} = \frac{a}{2}\begin{pmatrix} 1 \\ 1 \\ -1 \end{pmatrix}. \tag{6.3}$$

The body-centered unit cell comprises one formula unit. In order to represent the correct shape of the crystal potential in large voids additional augmentation spheres have to be inserted into the open crystal structure. The optimal augmentation sphere positions and radii of the spheres were determined automatically by the sphere geometry optimization algorithm [59]. The selected radii of the physical spheres and the calculated LDA valence charges are given in table 6.4. It is sufficient to dispose 48 empty spheres from 3 crystallographically inequivalent classes to keep the linear overlap of the real spheres below 15% and the overlap of any pair of real and empty spheres below 20%. Summing up, the unit cells of the ruthenates entering the band structure calculation consist of 68 spheres. The basis sets taken into account in the secular matrix contain Cu $4s$, $4p$, $3d$, $(4f)$, Ru $5s$, $5p$, $4d$, $4f$, $(3d)$, and O $2s$, $2p$, $(3d)$ orbitals as well as states of the additional augmentation spheres. Regarding the empty spheres the electronic configurations $1s$, $(2p)$, and $1s$, $2p$, $(3d)$ are applied. In addition, Na $3s$, $3p$, $3d$, $(4f)$, Ca $4s$, $4p$, $3d$, $4f$, $(5g)$, Sr $5s$, $5p$, $4d$, $4f$, $(5g)$, and La/Nd $6s$, $6p$, $5d$, $4f$, $(5g)$ states have to be considered for the A cations. States denoted in parentheses enter the calculation only as tails of other orbitals. During the course of the LDA calculations the Brillouin zones were sampled using an increasing number of **k**-points to ensure convergence of the results with respect to the fineness of the **k**-space grid. For each compound the number of **k**-points within the irreducible wedge of the Brillouin zone was 17, 34, 97, 212, 395, and finally 1255. The standard convergence criteria for the self-consistency were used. We address the question of chemical bonding, in addition to the study of partial densities of states, in the following section. Therefore we calculate the covalent bond energy as introduced at the end of chapter 2.

6.2 Results and Discussion

In Figure 6.5 the band structure as well as the partial Cu $3d$, Ru $4d$, and O $2p$ densities of states for the sodium compound $NaCu_3Ru_4O_{12}$ are displayed. Here the representation of the band structure refers to the first Brillouin zone of the body-centered cubic lattice, compare figure 6.2. One observes electronic bands in a wide energy range extending from about $-8.3\,\text{eV}$ to $0.7\,\text{eV}$. All bands reveal a considerable dispersion except for those located between $-3.0\,\text{eV}$ and $-2.0\,\text{eV}$. The electronic structure in the energy window shown is completely dominated by Cu $3d$, Ru $4d$, and O $2p$ states. In particular, contributions due to sodium are negligible small and therefore not included in figure 6.5. In the representation of the DOS three energy ranges can be distinguished. The broad interval from $-8.3\,\text{eV}$ to $-3.0\,\text{eV}$ is dominated by O $2p$ states. However, the oxygen contributions are complemented by noticeable admixtures from the transition metal d states due to covalent bonding. Within the energy range from $-3.0\,\text{eV}$ to $-2.0\,\text{eV}$ a sharp Cu $3d$ structure dominates, corresponding to the above mentioned less dispersing bands. In addition, the electronic states between $-2.0\,\text{eV}$ and $0.7\,\text{eV}$ and consequently the metallic conductivity

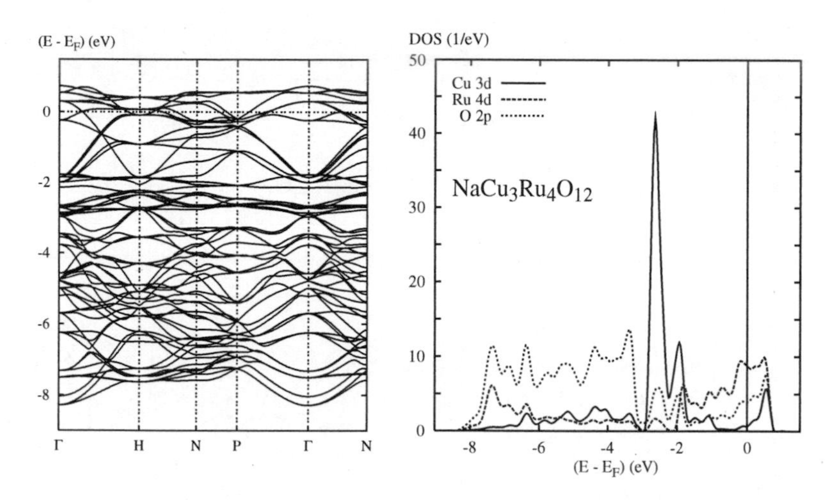

Figure 6.5: *Electronic bands (left) and partial densities of states (DOS) per formular unit (right) of* $NaCu_3Ru_4O_{12}$. *The bands are shown along selected symmetry lines in the first Brillouin zone of the body-centered cubic lattice. Figure 6.2 defines the symmetry points.*

of $NaCu_3Ru_4O_{12}$ derive mainly from broad ruthenium 4*d* bands. Covalent-type bonding yields finite O 2*p* contributions above −3.0 eV. The gross features of the band structures as well as the Cu 3*d*, Ru 4*d*, and O 2*p* densities of states are similar for all the studied compounds A=Na, Ca, Sr, La, Nd, see figures 6.5 and 6.6. As usual, contributions due to other than the displayed Cu 3*d*, Ru 4*d*, and O 2*p* orbitals are negligible in the region shown. In the neodymium compound additional Nd 4*f* states cause a sharp peak at and slightly above the Fermi level.

Covalent bonding between transition metal *d* and O 2*p* orbitals is confirmed by covalent bond energy curves calculated for the sodium compound as shown in figure 6.7. Qualitatively equivalent findings arise in the cases of the other compounds. Both the Cu-O and the Ru-O curves are negative and positive at energies below and above roughly −3.0 eV, respectively. Negative values are indicative of bonding states, whereas positive bond energies point to antibonding states. Bonding and antibonding contributions are observed in those energy regions where the O 2*p* and the transition metal *d* states dominate, respectively. Because of the higher *d* electron count of copper compared to ruthenium (for values see table 6.4) the Cu-O antibonding states are almost completely filled, yielding a vanishing contribution to the overall chemical bonding. Accordingly, the integrated Cu-O covalence energy curve from figure 6.7 crosses the Fermi energy close to zero. In contrast, the integrated Ru-O covalence energy takes a finite negative value at the Fermi level indicative of a stabilizing net contribution. Moreover, metal-metal bonding results in small Cu-Ru peaks between −3.0 eV and −2.0 eV as well as at 0.5 eV. Apart from the Nd 4*f* electrons all states not addressed in figures 6.5 and 6.6 play a secondary role close to the Fermi energy. In particular, electrons of the A cation do not contribute to covalent-type

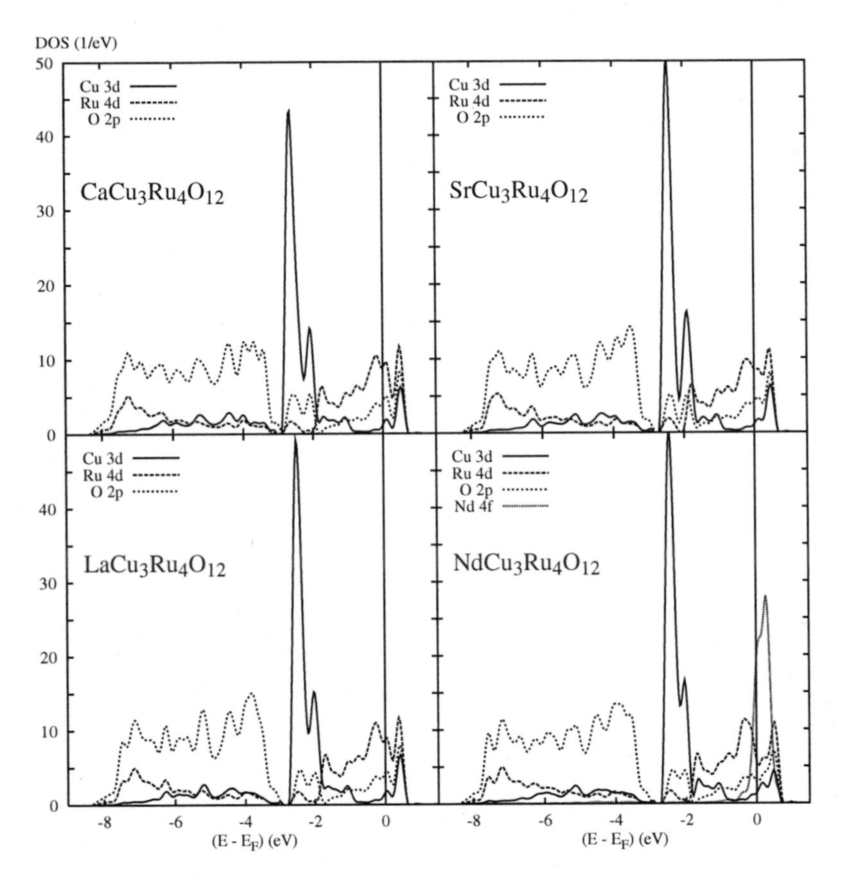

Figure 6.6: *Partial Cu 3d, Ru 4d, and O 2p densities of states (DOS) per formula unit of $ACu_3Ru_4O_{12}$ (A=Ca, Sr, La, Nd). For all compounds the findings are similar and coincide with figure 6.5. For $NdCu_3Ru_4O_{12}$ additional Nd 4f states appear near the Fermi energy.*

bonding but are distributed over the solid. Changes at the A site thus show up as slight modifications of the electron count but hardly influence the shape of the DOS.
S. G. Ebbinghaus *et al.* [135] related their measured atomic distances for the compounds $ACu_3Ru_4O_{12}$ (A=Na, Ca, Sr, La, Nd) to predictions of the bond valence model, which is an empirical approach connecting valences to ionic distances [136]. As it starts with ionic configurations it is less useful for covalently bonded crystals. The materials under study reveal the (formal) charge configurations $Na^{1+}Cu_3^{2+}Ru_4^{4.25+}O_{12}^{2-}$, $(Ca/Sr)^{2+}Cu_3^{2+}Ru_4^{4+}O_{12}^{2-}$, and $(La/Nd)^{3+}Cu_3^{2+}Ru_4^{3.75+}O_{12}^{2-}$. Applying ionic charges the bond valence approach leads to good agreements of the experimental and the calculated Ru-O bond lengths for A=Ca,

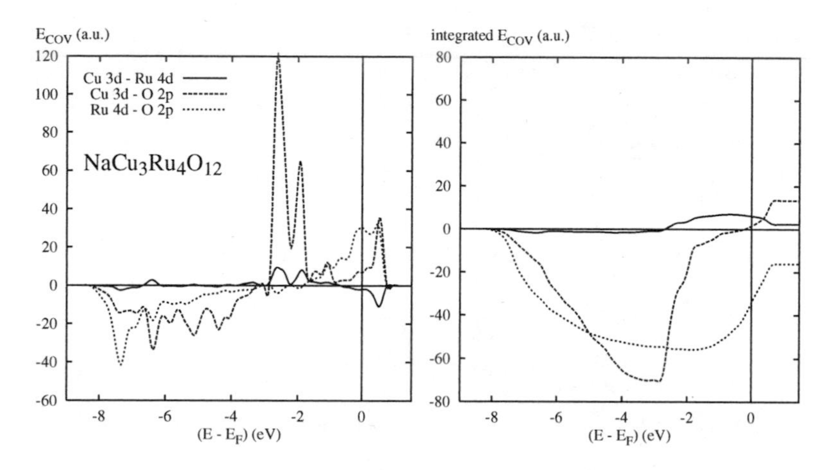

Figure 6.7: *Covalent bond energy for $NaCu_3Ru_4O_{12}$. Negative and positive values point to bonding and antibonding states, respectively. Integrated curves are shown on the right side.*

Sr (Ru^{4+}). In the A=Na ($Ru^{4.25+}$) case the measured Ru-O bond length is slightly larger than predicted by the bond valence theory. In contrast, it is somewhat shorter for A=La, Nd ($Ru^{3.75+}$). The experimental Cu^{2+}-O bond length within the CuO_4-square is slightly larger than predicted in the cases A=Na, Ca and significantly larger for A=Sr, La, Nd. Even more severe deviations from the bond valence results appear for the A-O bonds. In all compounds the measured distances are much too small. Going back to the calculated LDA valence charges in table 6.4 we find similar values for the Cu $3d$, Ru $4d$, and O $2p$ orbitals, independent of the particular A cation. Due to substantial covalent contributions to the bonding the calculated valences must deviate from the formal numbers, which also applies for the A cation. The accuracy of comparing calculated valences is limited since the assignment of electronic charge to specific atoms is arbitrary to some degree. In the present calculation the charge located within an atomic sphere is assigned to the respective atomic site. Electrons occupying the interstitial empty spheres cannot be reasonably assigned to any specific atom. As shown in table 6.4, the radii of corresponding spheres are very similar for all the compounds $ACu_3Ru_4O_{12}$. Of course, this fact makes a comparison of the calculated valences more reliable and we may state the following results: with only minor deviations between the five compounds one must deal with Cu^{2+} ions. Ruthenium appears as $Ru^{2.5+}$ in the sodium case and as $Ru^{2.4+}$ otherwise. The A cations realize the valences $Na^{0.9+}$, $Ca^{1.4+}$, $Sr^{1.3+}$, $La^{1.8+}$, and $Nd^{1.5+}$, which reflects strong deviations from an ionic picture and emphasizes the importance of covalent bonding.
In order to understand the consequences of the octahedral tilting for the electronic properties hypothetical crystal structures are investigated subsequently. Moreover, this allows for an analysis of the reported discrepancies between the measured and the expected interatomic distances. Since the specific choice of the A cation hardly influences the results we can concentrate on the sodium compound $NaCu_3Ru_4O_{12}$. LDA calculations for hypo-

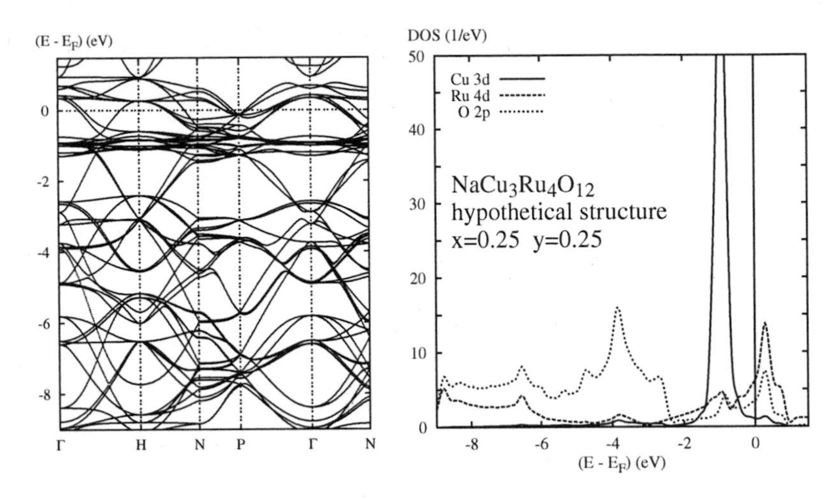

Figure 6.8: *Electronic bands (left) and partial densities of states (DOS) per formular unit (right) of hypothetically distorted $NaCu_3Ru_4O_{12}$. As usual, bands are shown along selected symmetry lines in the first Brillouin zone of the body-centered cubic lattice – as depicted in figure 6.2. The tilting of the oxygen octahedra as present in the real structure was omitted.*

thetical crystal structures offer the chance to study the interplay between the particular structural features and the resulting chemical stability. Here we will apply the following setups: we first shift the oxygen atoms away from the experimental positions back to the ideal perovskite positions given by the parameters $x_O = 0.25$ and $y_O = 0.25$. Second, the deviations from the ideal perovskite configuration are increased applying the parameters $x_O = 0.15$ and $y_O = 0.35$. These hypothetical configurations correspond to zero tilting of the octahedra as well as to an exaggerated tilting. Shifting the oxygen atoms in the above manner mainly affects the Cu-O bond lengths, whereas the effect for both the Ru-O and A-O distances is of second order. Hence the tilting barely influences the geometry of the oxygen octahedra. In addition, the octahedra keep a more or less constant distance from the A cations, although the oxygen atoms approach copper sites, see figure 6.3.
Band structures as well as partial densities of states for the hypothetical $NaCu_3Ru_4O_{12}$ crystal structures are displayed in figures 6.8 and 6.9. Substantial changes compared to the results for the experimental structure (figure 6.5) are observed. Turning to the ideal configuration the less dispersing copper-type bands are shifted to approximately $-1.0\,\mathrm{eV}$. The bands at lower energies show an increased dispersion in comparison to the results for the real crystal structure. One observes a considerable broadening of the O $2p$ dominated energy region, whereas the Cu $3d$ DOS sharpens strikingly. Both effects are readily understood from the oxygen shifts. On the one hand the latter lead to a strong increase of the Cu-O bond lengths from $1.94\,\text{Å}$ to $2.61\,\text{Å}$. Hence the Cu-O bonding almost vanishes, which leaves the copper atoms without relevant overlap to adjacent atoms and thus gives rise to a sharp DOS peak. On the other hand the Ru-O distances are somewhat smaller

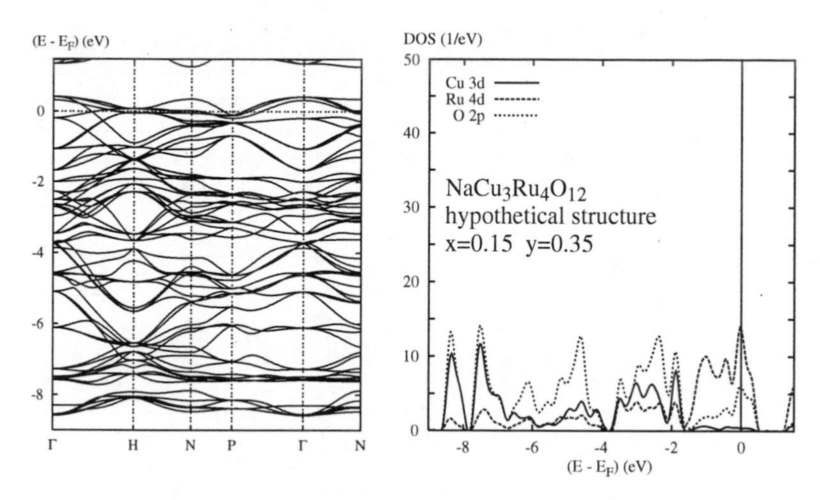

Figure 6.9: *Electronic bands (left) and partial densities of states (DOS) per formular unit (right) of hypothetically distorted $NaCu_3Ru_4O_{12}$. As usual, bands are shown along selected symmetry lines in the first Brillouin zone of the body-centered cubic lattice – as depicted in figure 6.2. The octahedral tilting was exaggerated compared to the real crystal structure.*

in the ideal (1.84 Å) than in the experimental (1.96 Å) structure, which we easily understand from figure 6.3. Thus the overlap of ruthenium $4d$ and oxygen $2p$ orbitals becomes slightly larger. As a consequence, larger Ru $4d$ contributions at energies below -3.0 eV are observed, whereas the Cu $3d$ fraction is almost negligible in this range. Ru $4d$ and O $2p$ admixtures to the pronounced Cu $3d$ peak are likewise very small. To sum up, shifting the oxygen atoms to their ideal perovskite positions yields an effective decoupling of the compound into RuO_6 octahedra and single Cu atoms. These constituents are effectively separated and no longer form a stable solid.
For the hypothetical crystal structure with exaggerated oxygen shifts the results change substantially. The pronounced copper $3d$ DOS peak and accordingly the hardly dispersing copper-type bands disappear. Consequently the copper states attain a most itinerant character and the oxygen states simultaneously move to remarkable lower energies. The alterations affecting the copper orbitals induce an increasing similarity of the Cu $3d$ and the O $2p$ partial densities of states due to larger covalent-type Cu-O bonding. This fact can be understood from the Cu-O distance amounting to only 1.57 Å in the exaggerated tilted structure. The fairly strong Cu-O bonding is also reflected by huge contributions to the covalent bond energy (not shown). Those are negative and positive below and above roughly -4.2 eV, respectively, leading to a net antibonding integral at the Fermi energy. While the bonding between Cu $3d$ and O $2p$ states increases, we observe a small decrease of the Ru-O bonding because the oxygen octahedra inflate slightly. As the Ru-O bonds are 2.12 Å long in the exaggeratedly tilted case, the integral of the covalence energy is less negative at the Fermi energy than for the real structure. Summarizing, the second hypo-

thetical structure suffers from an increase of antibonding Cu-like states at the expense of bonding Ru-like states.
Together with the instability of the previously discussed non-tilted $NaCu_3Ru_4O_{12}$ configuration we conclude that the experimental crystal structure is realized due to an optimal balance of different bonds. In particular, the bonding between Cu 3*d* and O 2*p* as well as between Ru 4*d* and O 2*p* orbitals plays an important role. The balance of these bonds is less affected by the specific A cations as they do not take part in the covalent bonding but have lost their valence electrons. From the chemical bonding point of view there is consequently no means to optimize the A-O bond length. The A-O distance is affected by the octahedral tilting to a rather small degree, even less than the Ru-O bond length. While the realized A-O distance amounts to 2.64 Å, the corresponding values for the non-tilted and the exaggeratedly tilted case are 2.61 Å and 2.81 Å, respectively. The competition of diverse bond strengths present in the $ACu_3Ru_4O_{12}$ (A=Na, Ca, Sr, La, Nd) compounds explains the deviations of the measured interatomic distances from those expected due to the bond valence approach. By means of the octahedral tilting it is possible to optimize the second strongest (Cu-O) bonds at a minor expense of the Ru-O distance. Recall that the crystal gains its stability mainly from the overlap of Ru 4*d* and O 2*p* orbitals.
In conclusion, the electronic properties of the perovskite-related ruthenates $ACu_3Ru_4O_{12}$ are governed by strong covalent bonding between transition metal *d* and O 2*p* electrons. While the Ru-O bonds (via their bond lengths) influence primarily the size of the oxygen octahedra and therefore the cubic lattice constant, octahedral tilting is predominately a consequence of Cu-O bonding. Compared to the ideal perovskite structure the distortions of the characteristic RuO_6 octahedra only slightly affect the Ru-O bonding. In contrast, the Cu-O distances undergo serious changes eventually leading to a square-planar coordination of copper. Finally, the unusual small A-O distances and the failure of the bond valence approach trace back to the competition of diverse bond strengths. The response of the A-O distance to the tilting is small and the structure does not offer any other possibility to simultaneously optimize both the Cu-O and the A-O bonds. Octahedral tilting, as discussed for the $ACu_3Ru_4O_{12}$ class of compounds, arises from the delicate interplay of covalent bonding between different atomic species yielding optimized bond lengths for all kinds of atoms involved. This general principle exhibits a rather universal nature and therefore applies to a large variety of compounds with similar atomic arrangements.

Chapter 7

Magnetic Chain Compounds

Low-dimensional systems are known for revealing fascinating physical properties. In the last years a new structural family of (quasi) one-dimensional compounds with interesting magnetic properties was synthesized. These materials belong to the hexagonal perovskite-type oxides and crystallize in a trigonal structure. The most frequently investigated member of the class is the magnetic chain compound $Ca_3Co_2O_6$, attracting much interest due to a possible partially disordered antiferromagnetism. Recently, the electronic properties and the magnetic ordering in $Ca_3Co_2O_6$ were analyzed by means of LDA band structure calculations [137]. Starting with these findings we subsequently study the closely related materials Ca_3CoRhO_6 and Ca_3FeRhO_6 to gain new insight into the microscopic origin of the magnetic order as well as the role of the particular atoms incorporated in the crystal structure.

7.1 Structural Considerations

As already mentioned, the perovskite family of oxides attracts attention due to its great compositional flexibility. In addition, a lot of modifications of the ideal perovskite structure are found, for instance as a consequence of octahedral tilting. Compared to the cubic compounds, the hexagonal perovskites are based on a variant of the oxygen packing. Both the cubic and hexagonal perovskites are generated from the stacking of close-packed AO_3 layers and the subsequent filling of the emerging octahedral sites with B cations. An abc-type stacking gives rise to the cubic perovskites, whereas an ab-type stacking results in the hexagonal perovskite structure. In the latter configuration we observe infinite chains of face-sharing AO_6 octahedra running along the c-axis of the hexagonal unit cell. These chains, in turn, are separated from one another by chains of B atoms. To be specific, the crystal structure of interest here is a modification of the hexagonal perovskite structure. Starting from the AO_3 layers it emerges by substituting one A' atom for three (adjacent) oxygen atoms. Stacking such $A_3A'O_6$ layers and occupying the octahedral sites with B cations leads to a new family of compounds with the general formula $A_3A'BO_6$. The main characteristic of this class of materials is the presence of chains of alternating face-sharing octahedra and trigonal prisms along the c_{hex}-axis. While the B cations are located at the

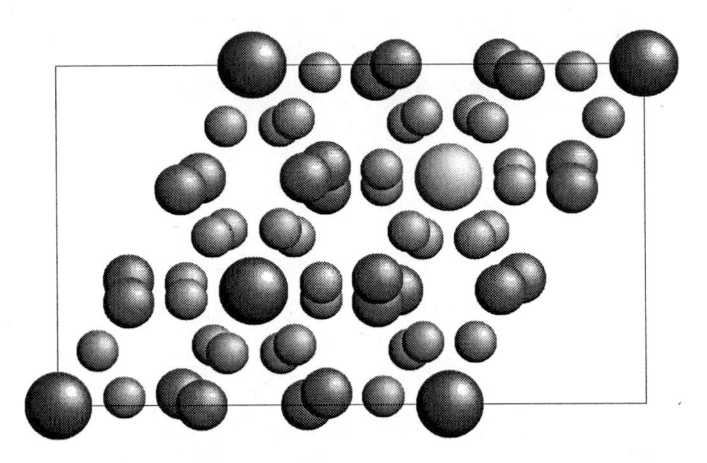

Figure 7.1: *Crystal structure of Ca_3CoRhO_6/Ca_3FeRhO_6 projected parallel to c_{hex}. Small and medium spheres represent oxygen and calcium, respectively. Large spheres denote both cobalt/iron (bright) and rhodium (dark), which alternate and set up chains parallel to c_{hex}.*

octahedral sites, the A′ cations occupy the trigonal prisms. As in the hexagonal perovskite structure, A cations separate the chains from each other. A trigonal prism nominally replaces two octahedra thus giving rise to a larger polyhedral site. In practice, the nominal size difference is reduced by a buckling of the oxygen layers. As the perovskite structure, the $A_3A'BO_6$ family accommodates many elements in a wide range of oxidation states. A summary of synthesized members has been given by K. E. Stitzer *et al.* [138].

In the following, we particularly discuss the compounds Ca_3CoRhO_6 and Ca_3FeRhO_6 and must now deal with face-sharing RhO_6 octahedra and CoO_6/FeO_6 trigonal prisms. These types of polyhedra alternate and consequently give rise to infinite chains along the c-axis of the (non-primitive) hexagonal unit cell. In figure 7.1 a projection of the crystal structure with the direction of view parallel to c_{hex} is presented. Both the polyhedral chains and the calcium atoms filling the interstitial regions are easily identified. The Co-Rh and Fe-Rh distances within the metal chains amount to 2.68 Å and 2.70 Å, respectively, which points to strong in-chain metal-metal bonding. Corresponding distances across the chains in each case are at least 5.61 Å. This value exceeds the 5.31 Å separation of neighbouring chains because the latter are shifted against each other parallel to the c_{hex}-axis by a third of the intrachain metal-metal distance. The structural arrangement becomes clearer from figure 7.2, which displays a second projection (parallel a_{hex}) of the crystal structure. The alternating arrangement of rhodium and cobalt/iron as well as the shift of neighbouring metal chains is visible. While rhodium occupies the octahedral sites, cobalt/iron is found at the centers of the trigonal prisms. To clarify the coordination of the metal sites some oxygen atoms are included in figure 7.2. Confirming the expectations due to the above discussion, the metal-oxygen distances in the octahedra are shorter than those in the trigonal prisms. In the cobalt/iron case the Co-O/Fe-O distance amounts to 2.15 Å/2.14 Å

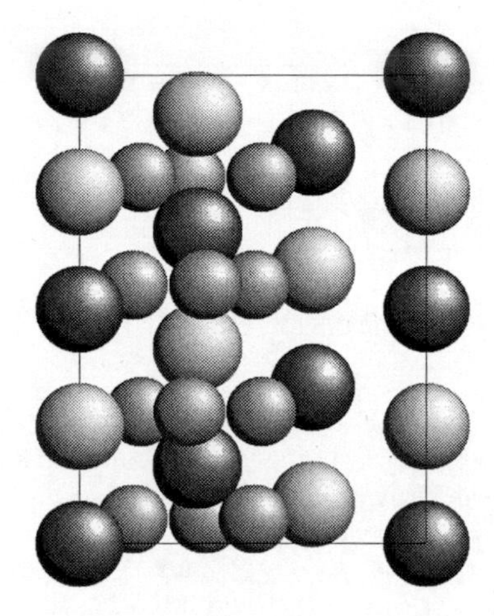

Figure 7.2: *Crystal structure of Ca_3CoRhO_6/Ca_3FeRhO_6 projected parallel to a_{hex}, using the same notation as in figure 7.1. All the calcium as well as most of the oxygen atoms are omitted. While rhodium occupies octahedral sites, cobalt/iron is located in trigonal prisms. Neighbouring chains are shifted parallel to c_{hex} by a third of the intrachain atomic distance.*

and the Rh-O distance to 2.00 Å/2.05 Å.

The structures of the compounds Ca_3CoRhO_6 as well as Ca_3FeRhO_6 have been determined by S. Niitaka *et al.* [139]. In both cases the room temperature powder x-ray diffraction data establish a trigonal lattice with space group $R\bar{3}c$ (D^6_{3d}). However, the authors use a non-primitive hexagonal unit cell containing six formula units instead of the two formula units the primitive trigonal cell comprises. The calcium atoms are found at the Wyckoff positions (18e): $\pm(x_{Ca}, 0, 1/4)$, $\pm(0, x_{Ca}, 1/4)$, $\pm(-x_{Ca}, -x_{Ca}, 1/4)$, $\pm(-1/3, 1/3, 1/3) \pm (x_{Ca}, 0, 1/4)$, $\pm(-1/3, 1/3, 1/3) \pm (0, x_{Ca}, 1/4)$, and $\pm(-1/3, 1/3, 1/3) \pm (-x_{Ca}, -x_{Ca}, 1/4)$. Cobalt/iron occupies the positions (6a): $\pm(0, 0, 1/4)$ and $\pm(-1/3, 1/3, 1/3) \pm (0, 0, 1/4)$. Rhodium atoms are observed at the positions (6b): $(0, 0, 0)$, $(0, 0, 1/2)$, $\pm(-1/3, 1/3, 1/3)$, and $\pm(-1/3, 1/3, 5/6)$. Finally, the oxygen atoms take the positions (36f): $\pm(x_O, y_O, z_O)$, $\pm(-y_O, x_O - y_O, z_O)$, $\pm(-x_O + y_O, -x_O, z_O)$, $\pm(y_O, x_O, 1/2 - z_O)$, $\pm(-x_O, -x_O + y_O, 1/2 - z_O)$, $\pm(x_O - y_O, -y_O, 1/2 - z_O)$, and sites shifted by $\pm(-1/3, 1/3, 1/3)$. In addition, the positional parameters are $x_{Ca} = 0.3666$, $x_O = 0.1823$, $y_O = 0.0230$, and $z_O = 0.1145$ in the case of Ca_3CoRhO_6. For Ca_3FeRhO_6 the corresponding values amount to $x_{Ca} = 0.3677$, $x_O = 0.1849$, $y_O = 0.0238$, and $z_O = 0.1181$. Furthermore, the hexagonal lattice constants are given by $a_H = 9.2017$ Å, $c_H = 10.7297$ Å and $a_H = 9.1960$ Å, $c_H = 10.7861$ Å in the cobalt and the iron case, respectively. Atomic coordinates in the hexagonal notation refer to the primitive translations defined in equation (3.4). For the band structure calculations we transform the hexagonal representation of the crystal lattice to the primitive trigonal representation. For this purpose we proceed analogous with the case of V_2O_3, where we were confronted with the same task.

	Ca_3CoRhO_6		Ca_3FeRhO_6	
Atom	Radius	Charge	Radius	Charge
Ca	2.6960	0.3572	2.7381	0.3884
Co/Fe	2.4933	7.2013	2.5444	6.0897
Rh	2.4718	6.7592	2.5470	6.9152
O	1.9119	4.1768	1.9701	4.2650

Table 7.3: *Atomic sphere radii (in a_B) and calculated LDA valence charges (Ca 3d, Co 3d, Fe 3d, Rh 4d, or O 2p) for ferromagnetic Ca_3CoRhO_6 and antiferromagnetic Ca_3FeRhO_6.*

Using equations (3.5) to (3.8) we set up the trigonal unit cell and transform each atomic position given in terms of the hexagonal primitive translations into the trigonal representation. We finally end up with the calcium atoms occupying the Wyckoff positions (6e): $\pm(x^*_{Ca}, 1/2 - x^*_{Ca}, 1/4)$, $\pm(1/4, x^*_{Ca}, 1/2 - x^*_{Ca})$, and $\pm(1/2 - x^*_{Ca}, 1/4, x^*_{Ca})$. Cobalt/iron takes the positions (2a): $\pm(1/4, 1/4, 1/4)$. Rhodium occupies the positions (2b): $(0, 0, 0)$ and $(1/2, 1/2, 1/2)$. Finally, the oxygen atoms are located at the Wyckoff positions (12f): (x^*_O, y^*_O, z^*_O), (z^*_O, x^*_O, y^*_O), (y^*_O, z^*_O, x^*_O), $(1/2 - y^*_O, 1/2 - x^*_O, 1/2 - z^*_O)$, $(1/2 - z^*_O, 1/2 - y^*_O, 1/2 - x^*_O)$, and $(1/2 - x^*_O, 1/2 - z^*_O, 1/2 - y^*_O)$. To complete the trigonal representation we state the positional parameters $x^*_{Ca} = x_{Ca} + 1/4 = 0.6166$, $x^*_O = 0.2968$, $y^*_O = 0.9552$, and $z^*_O = 0.0915$ in the case of Ca_3CoRhO_6. For Ca_3FeRhO_6 we have the values $x^*_{Ca} = x_{Ca} + 1/4 = 0.6177$, $x^*_O = 0.3030$, $y^*_O = 0.9570$, and $z^*_O = 0.0943$. The crystal structure of Ca_3FeRhO_6 additionally was characterized by means of single crystal x-ray diffraction [140]. However, the reported lattice constants and positional parameters are very similar to those stated before. Single crystal refinements of the structure of Ca_3CoRhO_6 are not published so far. For both the cobalt and the iron system data by S. Niitaka *et al.* [139] are used as structural input for the band structure calculations.
In the following, LDA results for non-magnetic and assumed ferromagnetic Ca_3CoRhO_6 as well as for non-magnetic and assumed antiferromagnetic Ca_3FeRhO_6 are discussed. In the non-magnetic calculations the spin degeneracy is enforced artificially. The unit cells comprise six structurally equivalent calcium, two equivalent cobalt/iron, two equivalent rhodium, and twelve equivalent oxygen atoms. In the case of Ca_3CoRhO_6 the magnetic coupling is assumed to be ferromagnetic both within the one-dimensional chains and between neighbouring chains. We subsequently call this kind of magnetic order the ferromagnetic configuration of Ca_3CoRhO_6. Furthermore, the magnetic structure of Ca_3FeRhO_6 is assumed to be antiferromagnetic within the transition metal chains as well as between neighbouring chains. This simply means the two crystallographically equivalent iron sites of the non-magnetic unit cell become magnetically inequivalent, which is likewise true for the two rhodium sites. To simplify matters, we will refer to the above magnetic order of Ca_3FeRhO_6 as the antiferromagnetic configuration.
In order to correctly model the crystal potential we insert additional augmentation spheres into the crystal structure. Optimal augmentation sphere positions and radii of all spheres were determined automatically by the sphere geometry optimization algorithm [59]. The selected radii of the physical spheres and the calculated LDA valence charges are summarized in table 7.3. In the rhodium case it is sufficient to dispose 66 additional spheres

out of 7 crystallographically inequivalent classes to keep the linear overlap of real spheres below 16% and the overlap of any pair of real and empty spheres below 21%. Thus the unit cell entering the Ca_3CoRhO_6 band structure calculation contains 88 spheres. For the iron compound 54 empty spheres out of 6 crystallographically inequivalent classes allow for overlap of real spheres of less than 17% and for overlap of any pair of real and empty spheres of less than 22%. Summing up, the unit cell of Ca_3FeRhO_6 comprises 76 spheres. The basis sets taken into consideration in the secular matrix consist of Ca $4s$, $4p$, $3d$, $(4f)$, Co/Fe $4s$, $4p$, $3d$, $(4f)$, Rh $5s$, $5p$, $4d$, $4f$, $(5g)$, and O $2s$, $2p$, $(3d)$ orbitals. Furthermore, states belonging to the additional augmentation spheres have to be included. Regarding the latter the electronic configurations $1s$, $(2p)$ and $1s$, $2p$, $(3d)$ were applied. As usual, states given in parentheses enter the LDA calculation as tails of other orbitals. In order to ensure convergence of the results with respect to the fineness of the **k**-space grid the Brillouin zones were sampled using an increasing amount of **k**-points (28, 60, 182, 408 to finally 770 points). Self-consistency of the charge density was accepted for deviations of the atomic charges and the total energy of subsequent iterations less than 10^{-8} electrons and 10^{-8} Ryd, respectively.

7.2 Ferromagnetic Coupling in Ca_3CoRhO_6

The magnetic chain compound Ca_3CoRhO_6 has started to receive attention due to unusual magnetic phase transitions it undergoes as a function of temperature. Importantly, the crystal structure is characterized by one-dimensional chains of alternating face-sharing RhO_6 octahedra and CoO_6 trigonal prisms. The magnetic properties appear to be strongly connected to this structural anisotropy. Note that the transition metal chains and therefore the magnetic ions form a triangular lattice. Magnetic susceptibility as well as neutron diffraction data reveal two magnetic transitions, the first at roughly 90 K and the second around 30 K [139, 141]. In the ordered state the magnetic ions couple ferromagnetically along the chains, while the interchain nearest neighbour interaction is antiferromagnetic. In particular, the interchain coupling causes magnetic frustration within the triangular arrangement. In a simple picture, only two thirds of the ferromagnetic chains may realize an antiferromagnetic order due to the geometrical frustration, whereas the remaining third stays incoherent. To be more specific, the incoherent chain is located at the center of a hexagon of antiferromagnetically coupled chains.
Up to now the magnetic properties of Ca_3CoRhO_6 are similar to those of $Ca_3Co_2O_6$ [142]. Admittedly, at a temperature of roughly 30 K the rhodium compound exhibits a second magnetic transition into a phase characterized by the freezing of the incoherent ferromagnetic chains. The randomly frozen spin chains show a much lower magnetic susceptibility than the fluctuating chains above 30 K. As in the case of $Ca_3Co_2O_6$ it seems reasonable to describe the magnetism of Ca_3CoRhO_6 in terms of a planar Ising triangular lattice where the magnetic moment of each chain plays the role of a single spin. While the incoherent Ising spins fluctuate at higher temperatures, they freeze in a random position at the 30 K transition. The response of the non-frozen incoherent ferromagnetic chains to an external field is very slow [143]. In full analogy to the findings for $Ca_3Co_2O_6$ a ferrimagnetic phase of Ca_3CoRhO_6 is stabilized applying a small magnetic field [144]. To be more spe-

cific, ferromagnetically coupled chains form a hexagonal network, to which the chains at the centers of the hexagons couple antiferromagnetically. Furthermore, sufficiently high magnetic fields force all chains to couple ferromagnetically. Recent ac susceptibility measurements show a peculiar behaviour in the temperature range from 40 K to 70 K with an unusually high frequency dependence, which points to an even more complex magnetic behaviour [146]. For the disordered phase of Ca_3CoRhO_6 neutron diffraction experiments of S. Niitaka *et al.* [141] suggest trivalent cobalt and rhodium ions. The trigonal prismatic cobalt atoms are observed in the high spin state $S = 2$, whereas the octahedral rhodium atoms reveal the low spin state $S = 0$. Hence the arrangement of the magnetic moments is exactly the same as in $Ca_3Co_2O_6$. The above magnetic structure is confirmed by the neutron diffraction study of M. Loewenhaupt *et al.* [147], who obtained a magnetic moment of $3.7\,\mu_B$ per ordered cobalt ion parallel to the c_{hex}-axis. No moment was found for the rhodium sites. Besides the antiferromagnetic long range order an additional magnetic short range order is observed, which affects the incoherent chains. These intrachain interactions seem isolated from the long range order as they persist also in the paramagnetic phase above 90 K. Altogether, Ca_3CoRhO_6 is characterized by a rich magnetic phase diagram closely related to that reported for $Ca_3Co_2O_6$. In the case of the latter compound the complex magnetic behaviour is complemented by peculiar transport properties [145]. Unfortunately, transport data for Ca_3CoRhO_6 are not available in the literature.
Due to very close relations between the magnetic features of $Ca_3Co_2O_6$ and Ca_3CoRhO_6 it is reasonable to assume common mechanisms of the phase transitions. For the former compound an investigation of the spin exchange interaction between cobalt atoms in high spin states ($3d^6$ configuration) resulted in ferromagnetic coupling within the chains [148]. Below we denote the octahedrally coordinated cobalt ions as Co1 and those in the trigonal prisms as Co2. A variation of the oxygen environment yields different crystal field splittings. In the case of the Co1 site there is the familiar splitting in lower t_{2g} and higher e_g states. Admittedly, the trigonal prismatic environment of Co2 accounts for another set of energy levels. In an atomic-like picture, this kind of local symmetry yields a scheme with the $d_{3z^2-r^2}$ state as the energetically lowest level followed by two pairs of twofold degenerate levels: d_{xy}, $d_{x^2-y^2}$ and d_{xz}, d_{yz}. Assuming the oxidation state 3+ for both cobalt sites we end with the electronic configuration $(t_{2g})^6$ and consequently without unpaired spin in the Co1 case. For the Co2 site a configuration with two electrons in the $d_{3z^2-r^2}$ orbital and two electrons in both the d_{xy,x^2-y^2} and the $d_{xz,yz}$ orbital allows for a high spin state with four unpaired spins and consequently a magnetic moment of $4\,\mu_B$ per formular unit as found experimentally. Spin polarized electronic band structure calculations performed by M.-H. Whangbo *et al.* [149] confirm both the oxidation state 3+ and the above distribution of unpaired spins. In the case of Ca_3CoRhO_6 these authors report qualitatively similar results. R. Frésard *et al.* [150] established an effective magnetic Hamiltonian for $Ca_3Co_2O_6$. For sufficiently strong Hund's rule coupling on the cobalt ions a ferromagnetic alignment of the $S = 2$ spins at the Co2 sites is stabilized, where the exchange is mediated by the $S = 0$ ions (Co1 sites). A further detailed analysis of the interchain magnetic interactions yields an antiferromagnetic coupling by means of super-superexchange via short O-O bonds, which is supported by LDA electronic structure calculations [137]. The same microscopic mechanism is expected to apply to other isostructural compounds, in particular to Ca_3CoRhO_6.

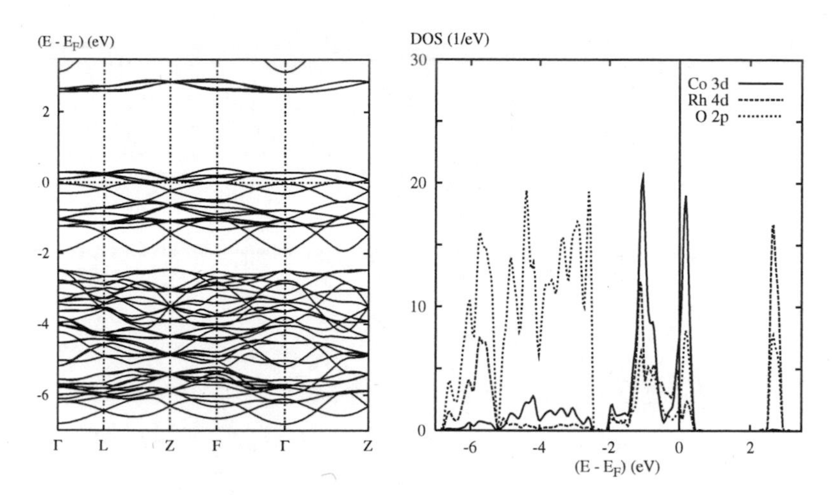

Figure 7.4: *Electronic bands (left) as well as partial densities of states (DOS) per unit cell of non-magnetic (spin degeneracy was enforced)* Ca_3CoRhO_6*. The bands are shown along selected symmetry lines in the first Brillouin zone of the trigonal lattice (see figure 3.12).*

For practical reasons we investigate the electronic and magnetic properties of Ca_3CoRhO_6 in two steps. First, we discuss a set of calculations where spin degeneracy was enforced. This allows us to address general issues such as the anisotropy of the electronic states, the crystal field splitting, the hybridization between transition metal and oxygen orbitals, or the chemical bonding. Afterwards we analyze an assumed ferromagnetic configuration to account for intrachain magnetic interactions. The magnetic properties of Ca_3CoRhO_6 will turn out to be most similar to those reported for $Ca_3Co_2O_6$ [137]. In the latter investigation also the antiferromagnetic interchain interaction was studied. However, the resulting electronic properties resemble the findings for the presumed ferromagnetic configuration. Thus we can skip an explicit investigation of the antiferromagnetic interaction in the case of Ca_3CoRhO_6 but transfer the previous $Ca_3Co_2O_6$ results.
The electronic band structure arising from the spin degenerate Ca_3CoRhO_6 LDA calculation is displayed in figure 7.4. Here the bands are depicted along selected high symmetry lines in the first Brillouin zone of the trigonal (rhombohedral) lattice, see figure 3.12. We have three groups of bands in the energy regions from $-6.8\,\text{eV}$ to $-2.4\,\text{eV}$, from $-2.0\,\text{eV}$ to $0.4\,\text{eV}$, and from $2.4\,\text{eV}$ to $3.0\,\text{eV}$. While the energetically lowest group consists of 36 individual bands, the middle and the highest group contain 16 and 4 bands, respectively. Next to the band structure in figure 7.4 corresponding Co $3d$, Rh $4d$, and O $2p$ densities of states are given in the same energy range. Contributions due to other than the shown orbitals can be neglected. The three aforementioned groups of bands naturally reappear as three distinct structures in the DOS. Obviously, the states in the energy interval from $-6.8\,\text{eV}$ to $-2.4\,\text{eV}$ mainly trace back to O $2p$ states. Since there are twelve oxygen atoms in each unit cell, the number of $12 \times 3 = 36$ electronic bands contained in this group is

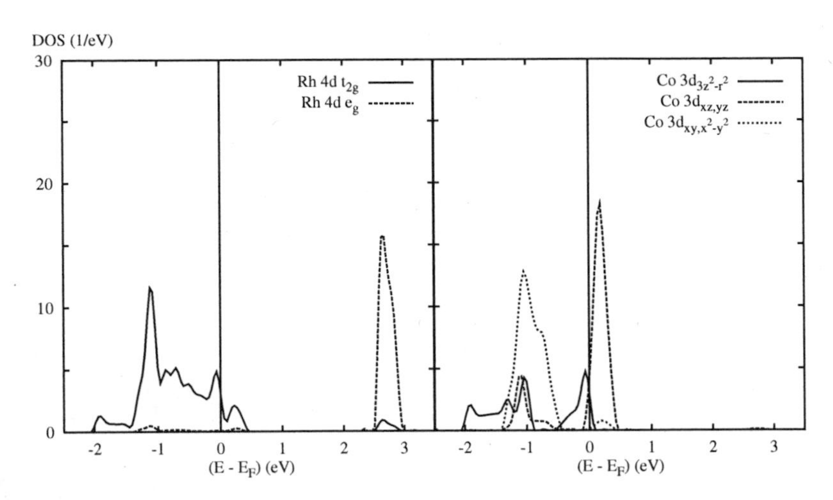

Figure 7.5: *Partial Rh 4d as well as Co 3d densities of states (DOS) per unit cell of non-magnetic Ca_3CoRhO_6. While the partial density at the octahedral rhodium sites separates into 4d t_{2g} and e_g states (with respect to the local rotated reference frame; left), the trigonal prismatic cobalt sites give rise to the components $3d_{3z^2-r^2}$, $3d_{xz,yz}$, and $3d_{xy,x^2-y^2}$ (right).*

understandable. The rhodium states are affected by the octahedral crystal field splitting into threefold degenerate $4d$ t_{2g} and twofold degenerate $4d$ e_g states, while the prismatic cobalt atoms are subject to the $3d$ splitting discussed earlier in the case of the Co2 sites of $Ca_3Co_2O_6$. Hence we interpret the Rh $4d$ dominated (highest) group of bands in figure 7.4 as the Rh $4d$ e_g states. Confirming our expectations, this group comprises $2 \times 2 = 4$ bands because of two rhodium atoms per unit cell. The $3 \times 2 = 6$ bands due to the Rh $4d$ t_{2g} states can be found in the energy interval from $-2.0\,\text{eV}$ to $0.4\,\text{eV}$. In this region we furthermore observe $5 \times 2 = 10$ bands originating from the Co $3d$ orbitals.
As it becomes obvious from the partial DOS shown in figure 7.4 there are substantial contributions of the transition metal $3d/4d$ and oxygen $2p$ states below and above $-2.0\,\text{eV}$, respectively. Admixtures in the energy regions dominated by the respective other states are a consequence of a remarkable p-d hybridization and reach up to roughly 30% of the total DOS. Contributions from the octahedral rhodium sites appear predominately from $-6.8\,\text{eV}$ to $-5.2\,\text{eV}$, hence setting up a broad structure. Co $3d$ admixtures are equally spread over the energy range from $-6.8\,\text{eV}$ to $-2.4\,\text{eV}$. Strong oxygen peaks accompany the transition metal peaks at $-1.1\,\text{eV}$, $0.2\,\text{eV}$, and $2.7\,\text{eV}$.
Crystal field splittings due to the octahedral and trigonal prismatic environments of the rhodium and cobalt atoms are found on closer inspection of the partial $4d$ and $3d$ DOS, respectively, see figure 7.5. Here the classification of the Rh $4d$ DOS in t_{2g} and e_g contributions refers to the local rotated reference frame established by the oxygen octahedra. As a consequence of the octahedral crystal field we note an almost perfect splitting into occupied t_{2g} and empty e_g states. In contrast, the trigonal prismatic crystal field at the

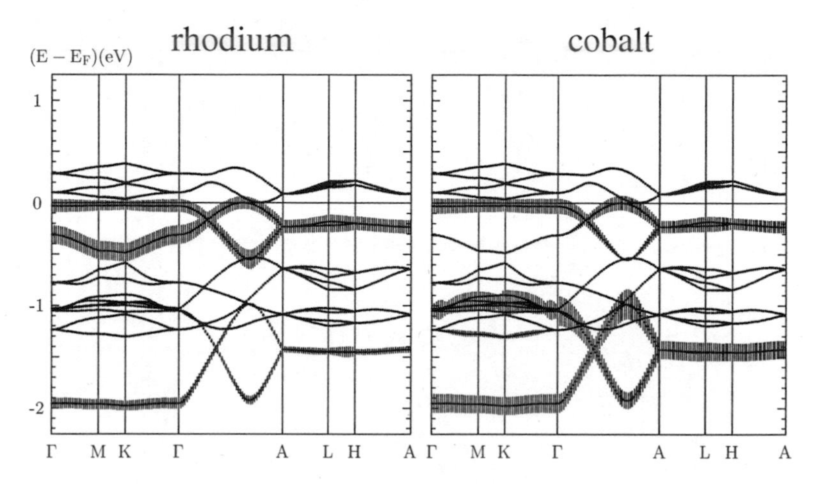

Figure 7.6: *Weighted electronic bands of non-magnetic Ca_3CoRhO_6 shown along selected symmetry lines in the first Brillouin zone of the hexagonal lattice, compare figure 3.6. The bars give the contributions due to the Rh 4$d_{3z^2-r^2}$ (left) and the Co 3$d_{3z^2-r^2}$ (right) orbital.*

cobalt atoms splits the 3d orbitals into non-degenerate $d_{3z^2-r^2}$ as well as twofold degenerate d_{xy,x^2-y^2} and $d_{xz,yz}$ states. While the latter two species mainly yield distinct DOS peaks around -1.0 eV and directly above the Fermi level, the $d_{3z^2-r^2}$ DOS is spread over a much wider energy range. In contrast to the predictions of the ideal (trigonal) crystal field splitting we thus find a remarkable portion of $d_{3z^2-r^2}$ states even above the d_{xy,x^2-y^2} peak. At the Fermi energy both $d_{3z^2-r^2}$ and $d_{xz,yz}$ states are present, whereas the d_{xy,x^2-y^2} contributions are very small. The broadening and splitting of the Co 3$d_{3z^2-r^2}$ DOS and the incomplete crystal field splitting indicate deviations from an ideal trigonal prismatic coordination, most likely due to intrachain metal-metal bonding. Almost all unoccupied states in figure 7.5 show Rh 4d e_g or Co 3$d_{xy,yz}$ character. Since the former orbitals mediate σ-type p-d overlap, the contributions to the DOS are found at much higher energies than the Co 3d e_g states.

In order to further study the shape of the Co 3$d_{3z^2-r^2}$ DOS and the electronic anisotropy induced by the quasi one-dimensional crystal structure, weighted band structures are displayed in figure 7.6. In the case of the left band structure the length of the bars added to the single bands represents the magnitude of the Rh 4$d_{3z^2-r^2}$ contributions to the different states. Importantly, the labeling here refers to the unrotated Cartesian coordinate system with the z-axis oriented along the Co-Rh chains. On the right hand side of figure 7.6 the Co 3$d_{3z^2-r^2}$ contributions are highlighted. In contrast to the band structure calculation presented in figure 7.4 we do not apply the first Brillouin zone of the primitive trigonal lattice in figure 7.6. Instead, these weighted bands refer to the non-primitive hexagonal representation as defined by equation (3.4) and the first Brillouin zone of the hexagonal lattice as illustrated in figure 3.12. This is most useful since the hexagonal symmetry lines

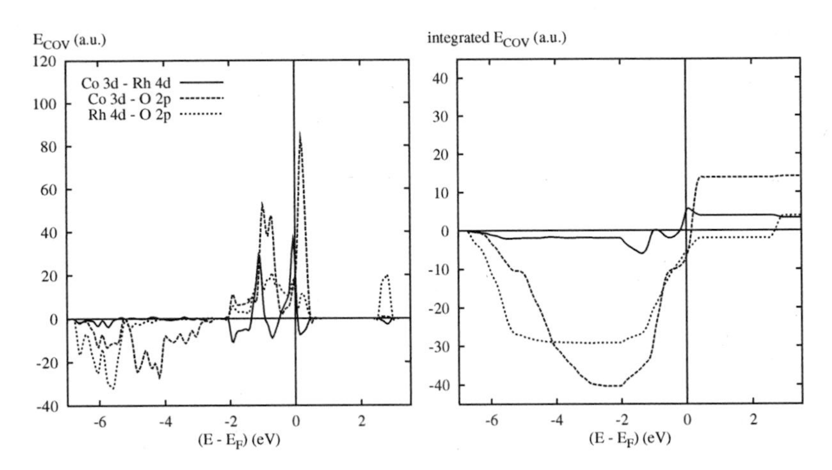

Figure 7.7: *Covalent bond energies of spin degenerate Ca_3CoRhO_6. Negative and positive values point to bonding and antibonding states, respectively. Integrated curves were added.*

reflect specific directions of the crystal lattice. In particular, the k_z-axis (Γ-A) represents the chain direction. As is obvious from figure 7.6 the $d_{3z^2-r^2}$-type bands show a considerable dispersion along the chains, while the dispersion perpendicular is almost negligible. Moreover, the in-chain dispersion (Γ-A) of the $d_{3z^2-r^2}$ bands is stronger as compared to the other bands, which reflects the one-dimensionality of the crystal structure.

According to the decomposition of the partial Co 3*d* DOS into its symmetry components the occupied bands without visible bars in figure 7.6 are influenced by the Co $3d_{xy,x^2-y^2}$ states. The bands directly above the Fermi energy trace back to the Co $3d_{xz,yz}$ orbitals. Both the $3d_{xy,x^2-y^2}$-like and $3d_{xz,yz}$-like states show an increased dispersion in the hexagonal planes, indicative of Co-O overlap and/or interchain coupling. In contrast, the Rh $4d_{3z^2-r^2}$ and Co $3d_{3z^2-r^2}$ orbitals mediate intrachain metal-metal overlap along c_{hex}. The contributions of the latter states to the bands directly below the Fermi energy and to the lowermost bands point to strong metal-metal bonding along the Co-Rh chains. This conclusion is consistent with large $d_{3z^2-r^2}$ band widths and eventually explains the peculiar shape of the Co $3d_{3z^2-r^2}$ DOS. While the bonding states are cobalt-like, the antibonding branch shows an increased rhodium character.

Strong hints at relevant metal-metal bonding are derived from the covalent bond energy curves given in figure 7.7. Negative and positive covalence energies again are indicative of bonding and antibonding states, respectively. Obviously, metal-oxygen overlap leads to bonding contributions below $-2.4\,\text{eV}$, whereas antibonding states are located at energies above $-2.0\,\text{eV}$. We have already discussed the increased Rh 4*d* DOS in the energy region from $-6.8\,\text{eV}$ to $-5.2\,\text{eV}$ because of hybridization effects (compare figure 7.4). Confirming this observation negative covalence energies, pointing to bonding between Rh 4*d* and O 2*p* orbitals, appear in the same energy range. Bonding contributions tracing back to the overlap of Co 3*d* and O 2*p* orbitals predominately are found from $-5.2\,\text{eV}$ to $-2.4\,\text{eV}$. This again confirms our expectations from the DOS. Metal-metal bonding shows up mainly in the upper region of the valence band. According to our previous band structure analysis the negative bond energy peak around $-1.9\,\mathrm{eV}$ as well as the positive peak near $-0.1\,\mathrm{eV}$ correspond to the transition metal $d_{3z^2-r^2}$ orbitals thus supporting a strong in-chain Co-Rh interaction. Those peaks in between are difficult to interpret since not only the $d_{3z^2-r^2}$ states but also all other d electrons contribute in this region. Finally, we study the chemical bonding in the Ca_3CoRhO_6 crystal by means of the integrated bond energy curves depicted on the right hand side of figure 7.7. While metal-metal overlap does not stabilize the crystal, the covalent bond energies representing metal-oxygen bonding cross the Fermi energy at similar negative values. Both the Co-O and the Rh-O bonds therefore yield a stabilizing net contribution.

Due to the characteristics of the crystal structure of Ca_3CoRhO_6, which are reflected by the anisotropy of the electronic states, it is reasonable to assume an increased magnetic exchange coupling within the Co-Rh chains as compared to a possible interchain coupling. We therefore investigate the magnetic properties of Ca_3CoRhO_6 by means of an assumed ferromagnetic order. Implications of the additional antiferromagnetic interchain coupling will be investigated afterwards. The spin polarized electronic band structure calculations performed by M.-H. Whangbo *et al.* [149] likewise assume a ferromagnetic order. In this respect the present analysis is along the same line of reasoning. In contrast to the goals of the above authors we do not primarily aim at an investigation of the atomic oxidation states or the spin distribution in Ca_3CoRhO_6, but study instead the microscopic origin of the magnetic couplings. Starting with the previously presented non-magnetic LDA calculation it is possible to obtain a converged solution with well localized magnetic moments (per atom): $0.48\,\mu_B$ for rhodium, $2.59\,\mu_B$ for cobalt, $0.14\,\mu_B$ for oxygen, and $0.00\,\mu_B$ for calcium. The magnetic moment of the unit cell sums up to $7.94\,\mu_B$. Apart from lifting the spin degeneracy, all technical details of the assumed ferromagnetic LDA calculation are the same as in the non-magnetic case.
The calculated magnetic moments of Ca_3CoRhO_6 are closely related to that reported for $Ca_3Co_2O_6$: $0.35\,\mu_B$ per Co1 atom, $2.73\,\mu_B$ per Co2 atom, $0.14\,\mu_B$ per O atom, $0.01\,\mu_B$ per Ca atom, and consequently $8.00\,\mu_B$ per unit cell [137]. Furthermore, they reflect the experimentally determined low and high spin states of the rhodium atoms at the octahedral and the cobalt atoms at the trigonal prismatic sites, respectively. Corresponding partial spin-majority and spin-minority densities of states as displayed in figure 7.8 confirm the small spin splitting at the rhodium sites. In contrast, the spin-majority and spin-minority Co $3d$ densities deviate considerably. While the shape of the spin-minority Co $3d$ DOS resembles the gross features of the non-magnetic DOS (see figure 7.5), the spin-minority states show strong shifts to lower energies. In particular in the upper region of the oxygen dominated energy range from approximately $-5.0\,\mathrm{eV}$ to $-1.8\,\mathrm{eV}$ remarkable Co $3d$ contributions can be observed. Admixtures by the transition metal states in the lower oxygen dominated energy region basically trace back to the Rh $4d$ electrons. They are very similar in the cases of the spin-majority, spin-minority, and non-magnetic DOS. Moreover, one easily observes O $2p$ contributions above $-1.8\,\mathrm{eV}$ accompanying the transition metal peaks in both diagrams of figure 7.8. As in the non-magnetic case, such admixtures in the energy regions dominated by the respective other states are due to p-d hybridization.

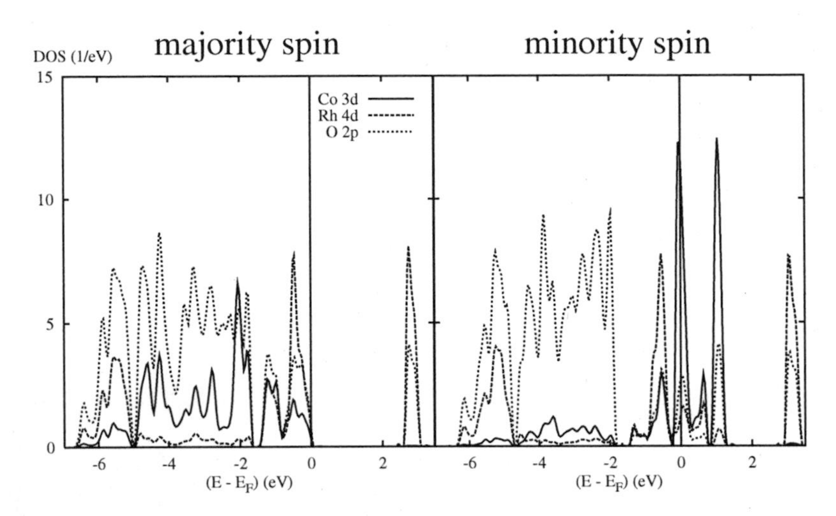

Figure 7.8: *Partial Co 3d, Rh 4d, and O 2p densities of states (DOS) per unit cell of ferromagnetic Ca_3CoRhO_6 – separated into the spin-majority and spin-minority contributions.*

Especially hybridization of Co $3d$ and O $2p$ orbitals may account for the rather high magnetic moment at the oxygen sites. A strong polarization of the ligand states is indicative of extended moment formation as found in different copper oxides [151, 152]. Extended moment formation here is defined as the combination of strong ligand moments with the fact that the ligands play a minor role for the exchange coupling. This situation is realized in the compound Ca_3CoRhO_6.

In order to analyze the extended moment formation in more detail we turn to figure 7.9 showing the partial Rh $4d$ as well as Co $3d$ DOS. The classification of the Rh $4d$ density in t_{2g} and e_g contributions again refers to the local rotated reference frame. For reasons of clarity both the spin-majority and the spin-minority densities of states are given in figure 7.9. Examinating the DOS curves reveals a similar halfmetallic behaviour as reported for assumed ferromagnetic $Ca_3Co_2O_6$. However, since the Rh $4d$ DOS of Ca_3CoRhO_6 is much broader than the corresponding Co $3d$ DOS of $Ca_3Co_2O_6$, the former material is less close to an insulating state than the latter. As in the spin degenerate case of Ca_3CoRhO_6 we find in figure 7.9 a strong crystal field splitting into $4d$ t_{2g} and e_g states at the octahedral rhodium sites. While the t_{2g} states appear in the vicinity of the Fermi level, the e_g states are found at high energies of roughly 2.5 eV. Because of the hybridization with the O $2p$ orbitals additional Rh $4d$ t_{2g} and e_g contributions occur around −5.5 eV. At the cobalt sites we distinguish between the non-degenerate $3d_{3z^2-r^2}$, twofold degenerate $3d_{xz,yz}$, and likewise twofold degenerate $3d_{xy,x^2-y^2}$ symmetry components. In contrast to the rhodium atoms the trigonal prismatic sites are subject to a reduced crystal field splitting. Due to hybridization with the O $2p$ orbitals the cobalt $3d$ spin-majority states hence shift below the Fermi level giving rise to a very broad DOS shape. Obviously, this behaviour is consis-

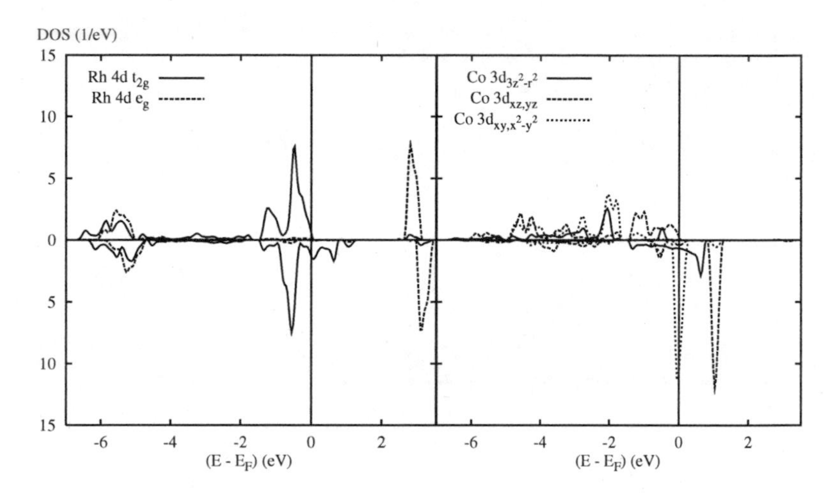

Figure 7.9: *Partial Rh 4d and Co 3d densities of states (DOS) per unit cell of ferromagnetic Ca_3CoRhO_6 – separated into spin-majority (top) and spin-minority (bottom) contributions. While the partial density of states due to the octahedral rhodium sites splits up into 4d t_{2g} and e_g states (with respect to the local rotated reference frame; left), the trigonal prismatic cobalt sites give rise to the components $3d_{3z^2-r^2}$, $3d_{xz,yz}$, and $3d_{xy,x^2-y^2}$ (right).*

tent with extended moment formation. Spin-minority $3d$ admixtures in the energy range dominated by the oxygen states are fairly small. As in the non-magnetic case the $3d_{xz,yz}$ and $3d_{xy,x^2-y^2}$ densities of states show single sharp maxima. Compared to figure 7.5, the peaks are settled at higher energies, thus giving only minor contributions to the occupied states. The spin-minority Co $3d_{3z^2-r^2}$ DOS is characterized by its broad shape stretching from about $-1.6\,\text{eV}$ to $0.8\,\text{eV}$. The finite DOS at the Fermi level clearly characterizes assumed ferromagnetic Ca_3CoRhO_6 as a metal. However, the resulting conductivity should be both anisotropic and spin dependent. Due to lack of investigation it remains an open question whether these results are in agreement with experimental data. Comparing the presumed ferromagnetic configuration of Ca_3CoRhO_6 to the enforced spin degenerate case the total energy per unit cell decreases by about 36 mRyd. This supports the experimental observation of an in-chain ferromagnetism. Because the interchain coupling is found to be much weaker, it should yield only minor contributions to the total energy.
Implications of the energetical downshift of the Co $3d$ spin-majority states become clearer from the energy and orbital dependent spin magnetizations depicted in figure 7.10. Such spin magnetizations are calculated from the DOS as the integrated difference of the spin-majority and spin-minority densities of states, see figure 7.9. In the case of the rhodium magnetizations, the separation into $4d$ t_{2g} and e_g contributions once more refers to the reference frame of the oxygen octahedra. In the left diagram of figure 7.10 additionally the $d_{3z^2-r^2}$ magnetization is depicted. Here the notation refers to the non-rotated coordinate system, which likewise holds for the cobalt symmetry components in the right hand

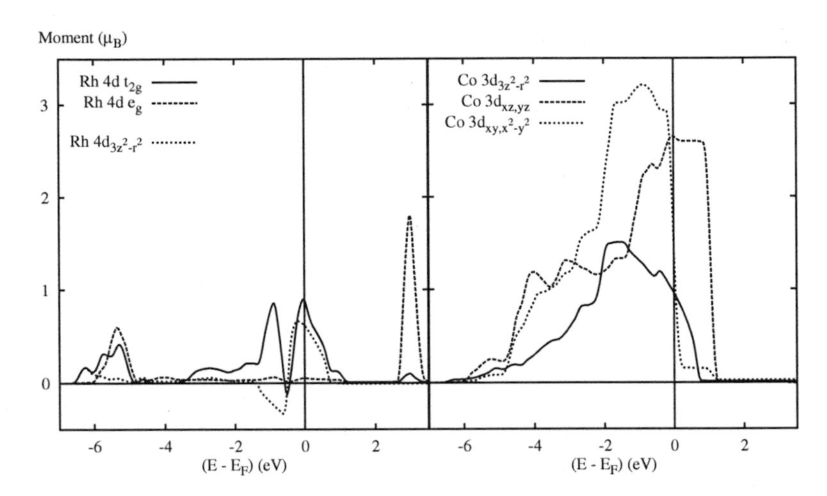

Figure 7.10: *Magnetic moments per unit cell of ferromagnetic Ca_3CoRhO_6 (Rh 4d and Co 3d). The notation is the same as introduced previously for the partial DOS (see figure 7.9).*

diagram. Recall that the principal axes of the $d_{3z^2-r^2}$ orbitals are oriented along the c_{hex}-axis, i.e. along the Co-Rh chains. According to figure 7.10 the small magnetic moment at the octahedrally coordinated rhodium sites is carried almost exclusively by the $4d$ t_{2g} orbitals. Because the spin splitting at these sites is small and the $4d$ e_g states are located well above the Fermi level they do not contribute to the net magnetization. The magnetic moment of the e_g orbitals hence crosses the Fermi level almost at zero, whereas the t_{2g} curve takes a finite value. Applying the non-rotated reference frame one can analyze the origin of the rhodium moment more exactly. Because the $d_{3z^2-r^2}$ orbital accounts for the main part of the t_{2g} magnetization the spin polarization seems to be connected to the metal-metal bonding in the direction of the Co-Rh chains.
Compared to the rhodium atoms, the cobalt sites are subject to a much larger spin splitting, which affects all the $3d$ states to a similar degree. Bear in mind that the $d_{xz,yz}$ and d_{xy,x^2-y^2} symmetry components in figure 7.10 each represent two of the atomic orbitals. Summing over the occupied states up to the Fermi level, we obtain a magnetic moment of about $0.50\,\mu_B$ carried by each $d_{3z^2-r^2}$ orbital. Moreover, the $d_{xz,yz}$ moment amounts to roughly $0.65\,\mu_B$ per orbital, whereas the d_{xy,x^2-y^2} moment is $0.40\,\mu_B$ per orbital. These values are almost identical to the numbers reported for $Ca_3Co_2O_6$. Finally, the extended moment formation at the cobalt sites is directly reflected by figure 7.10. Obviously, the moment starts to accumulate already at about $-6.0\,$eV, i.e. well inside the energy interval dominated by the O $2p$ states. This kind of moment accumulation is a consequence of the strong p-d hybridization already discussed in the context of the spin degenerate LDA calculation. Because of the hybridization the O $2p$ orbitals can be easily polarized, yielding a considerable magnetic moment of $0.14\,\mu_B$. The O $2p$ polarization as well as the extended moment formation at the trigonal prismatic sites are found in almost the same

manner in the case of $Ca_3Co_2O_6$.
Similar to the latter compound the ferromagnetic coupling in the transition metal chains of Ca_3CoRhO_6 seems to be triggered mainly by the atoms at the trigonal prismatic sites. Due to hybridization between Co $3d$ and O $2p$ orbitals combined isotropic moments are established. Because the O $2p$ orbitals participate in the local magnetic moment, but not in the spin exchange coupling between the trigonal prisms, the above behaviour reflects an extended moment formation. The coupling of the extended – but nevertheless well localized – moments is mediated by superexchange interaction via $d_{3z^3-r^2}$ orbitals. Strong σ-type metal-oxygen bonding in the RhO_6 octahedra allows for only small spin-splittings at the rhodium sites. Since all these results reflect (even quantitatively) the findings for $Ca_3Co_2O_6$ [137, 149] it is reasonable to conclude that the octahedrally coordinated atoms play only a secondary role for the in-chain magnetic moment, but are most important for the exchange coupling. Replacing cobalt by rhodium at the octahedral sites hardly affects the magnetic properties. Confirming the findings of C. Laschinger *et al.* [148] the in-chain ferromagnetism seems to arise from the spin exchange coupling between the cobalt atoms in the trigonal prisms. However, this coupling is mediated by the transition metal atoms located at the octahedral sites.
Replacing cobalt by rhodium does not modify the magnetic characteristics of the trigonal prismatic sites only because these atoms have strongly related electronic configurations. The formal $[\mathrm{Ar}]3d^7 4s^2$ set-up of cobalt leads to a similar electron count in the system as induced by the formal $[\mathrm{Kr}]4d^8 5s^1$ set-up of rhodium. Tabel 7.3 gives the calculated LDA valence charges for the presumed ferromagnetic in-chain coupling. As usual, we have to consider the limitations of assigning electronic charge to particular atomic sites. Here the radii of the different transition metal spheres are similar thus allowing for a more reliable comparison. Apart from the inevitable limitations we obtain the oxidation states $Rh^{2.24+}$ and $Co^{1.80+}$. Contradicting the results of M.-H. Whangbo *et al.* [149] we note a distinct difference of the valences at the octahedral and the trigonal prismatic sites. In addition, the valences differ remarkably from the ideal ionic value 3+, which is not surprising due to the covalent bonding present in Ca_3CoRhO_6. For this reason the simple ionic picture of the magnetism is questionable. Remember that the oxidation state 3+ leaves six valence electrons on the transition metal sites. In the rhodium case this leads to fully occupied t_{2g} orbitals and hence to an $S = 0$ low spin state. For the trigonal prismatically coordinated sites one expects two electrons in the $d_{3z^2-r^2}$ orbital and one electron in each of the four $d_{xy,x^2-y^2}/d_{xz,yz}$ orbitals, yielding an $S = 2$ high spin state. The LDA calculation finds a magnetic moment of $2.59\,\mu_B$ per cobalt site and confirms the experimentally determined value of slightly less than $4.00\,\mu_B$ per formula unit. Importantly, not the whole moment is found at the trigonal prismatic sites but, due to hybridization effects, the rhodium and oxygen atoms also contribute.
After the analysis of the ferromagnetic intrachain coupling of Ca_3CoRhO_6 we now turn to the antiferromagnetic interchain coupling. Due to almost identical electronic properties of Ca_3CoRhO_6 and $Ca_3Co_2O_6$ we can transfer the insights obtained in the latter case [137] to our present investigation. Concerning the single transition metal chain, a ferrimagnetic configuration of $Ca_3Co_2O_6$ gives rise to almost the same electronic features as observed in the ferromagnetic case. In particular, all the local magnetic moments are very similar in size. There are neither significant differences in the distribution of the moments between

the high and low spin transition metal sites nor in the polarization of the oxygen atoms. The small magnetic moments at the octahedral sites are carried essentially by the $3d$ t_{2g} states in a narrow energy region next to the Fermi energy. In contrast, the moments at the trigonal prismatic sites arise from similar contributions of all five d orbitals. In addition, by virtue of the p-d hybridization they already built up at low energies. In the ferrimagnetic arrangement two types of transition metal chains are distinguished since there are twice as many of the first as of the second chain. The calculated LDA densities of states of the majority and minority chains overall are most similar. However, one observes slightly sharper transition metal d peaks in the case of the minority chains. This fact points to an increased localization of the respective orbitals. Summarizing, the interchain coupling in the ferrimagnetic configuration of $Ca_3Co_2O_6$ influences the local electronic and magnetic properties to a minor degree. All these conclusions should likewise apply to Ca_3CoRhO_6. While the spin exchange coupling of the high spin cobalt atoms establishes the intrachain ferromagnetism, the mutual alignment of the single chains seems to be independently determined by the antiferromagnetic interchain coupling. Due to an unusually large Co-O hybridization super-superexchange via short O-O bonds may account for the latter [150]. In the disordered phase only two thirds of the transition metal chains can order because of the geometrical frustration. The application of an external field makes a ferrimagnetic or ferromagnetic arrangement favorable [144].

7.3 Antiferromagnetic Coupling in Ca_3FeRhO_6

The analysis of Ca_3CoRhO_6 in the preceding section revealed no qualitative modification of the electronic and magnetic properties caused by exchanging the octahedrally coordinated cobalt atoms in $Ca_3Co_2O_6$ for isoelectronic rhodium. In the following, we will investigate another isostructural material, which permits us to exchange the atoms located at the trigonal prismatic sites: Ca_3FeRhO_6. The crystal structure of this compound is dominated by one-dimensional chains of alternating face-sharing RhO_6 octahedra and $FeCo_6$ trigonal prisms. As in the previously discussed materials the electronic and magnetic features of Ca_3FeRhO_6 are strongly influenced by its structural anisotropy. Remarkably, the magnetism of Ca_3FeRhO_6 turns out to be much less complex than reported in the cases of $Ca_3Co_2O_6$ and Ca_3CoRhO_6. Replacing the magnetic cobalt atoms of Ca_3CoRhO_6 with iron and comparing the material properties allows for deeper insight into the microscopic mechanisms of the magnetic coupling.
When decreasing the temperature below 12 K magnetic susceptibility measurements for Ca_3FeRhO_6 indicate a transition from a paramagnetic into a three-dimensional antiferromagnetic phase [139, 140]. The easy axis of the ordered antiferromagnetic arrangement is oriented along the one-dimensional Fe-Rh chains. Furthermore, fitting of the susceptibility data suggests divalent iron with an $S = 2$ high spin state, while rhodium is found to be tetravalent with an $S = 1/2$ low spin state. According to ^{57}Fe Mössbauer experiments for the magnetically ordered phase, the iron atoms occupy crystallographically equivalent trigonal prismatic sites and show the same valence [154]. Contradicting the previous findings from the magnetic susceptibilities, a large hyperfine field is characteristic of trivalent iron in an $S = 5/2$ spin state. The saturation magnetization measured by applying strong

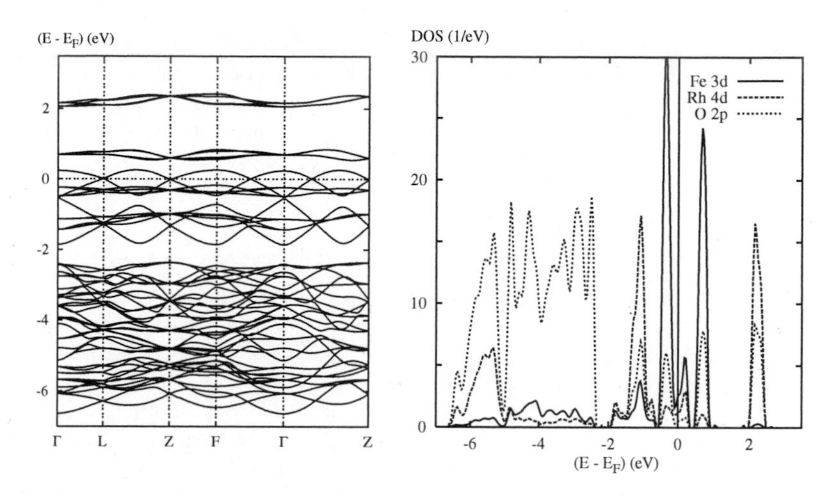

Figure 7.11: *Electronic bands (left) as well as partial densities of states (DOS) per unit cell of non-magnetic (spin degeneracy was enforced) Ca_3FeRhO_6. The bands are shown along selected symmetry lines in the first Brillouin zone of the trigonal lattice (see figure 3.12).*

pulsed external magnetic fields amounts to $3.74\,\mu_B$ per formula unit. This is considerably less than the $5.00\,\mu_B$ per formula unit expected for a configuration based on the valency stages Fe^{3+} ($S = 5/2$) and Rh^{3+} ($S = 0$). To overcome the contradictions arising from the susceptibility and the Mössbauer data it was suggested that a substantial part of the iron atoms at the trigonal prismatic sites is replaced by rhodium. However, for the LDA band structure calculation the ideal Ca_3FeRhO_6 configuration is used.

Analogous with our previous investigation of Ca_3CoRhO_6 we will study the electronic and magnetic properties of Ca_3FeRhO_6 in two steps. First, we discuss an LDA band structure calculation where spin degeneracy is enforced artificially to address more general features such as the electronic anisotropy induced by the crystal structure, crystal field splittings, possible hybridization effects, and chemical bonding. Second, we focus on the intrachain magnetic interactions by concentrating on the antiferromagnetic order. In particular, the findings are compared to results for the ferromagnetic in-chain coupling in Ca_3CoRhO_6. In spite of their diverse magnetic structures we will observe remarkable similarities in the electronic and magnetic properties of both compounds.
Enforced spin degenerate LDA results are presented in figure 7.11. The electronic bands displayed on the left hand side of the figure refer to selected high symmetry lines in the first Brillouin zone of the trigonal (rhombohedral) lattice, see figure 3.12. In full analogy to the previous Ca_3CoRhO_6 study we easily identify three groups of bands ranging from approximately $-6.7\,\text{eV}$ to $-2.3\,\text{eV}$, from $-2.1\,\text{eV}$ to $0.9\,\text{eV}$, and from $2.0\,\text{eV}$ to $2.4\,\text{eV}$. The lowest, the middle, and the highest group once more comprise 36, 16, and 4 bands, respectively. Thus we interpret the $12 \times 3 = 36$ bands of the lowest group as reminiscent

of the $2p$ orbitals of the twelve oxygen atoms in the Ca_3FeRhO_6 unit cell. The rhodium atoms are affected by the octahedral crystal field splitting leading to threefold degenerate $4d$ t_{2g} and twofold degenerate $4d$ e_g states. In contrast, the trigonal prismatically coordinated iron atoms are influenced by a crystal field splitting into non-degenerate $3d_{3z^2-r^2}$, twofold degenerate $3d_{xy,x^2-y^2}$, and likewise twofold degenerate $3d_{xz,yz}$ states. While the $2 \times 2 = 4$ bands in the energy range from 2.0 eV to 2.4 eV are reminiscent of the Rh $4d$ e_g states, the remaining $3 \times 2 = 6$ rhodium bands due to the t_{2g} manifold are observed in the interval from -2.1 eV to 0.9 eV. Here we also find $5 \times 2 = 10$ bands originating from the Fe $3d$ orbitals. Note the two iron and two rhodium atoms per unit cell.
In addition to the band structure, figure 7.11 depicts corresponding Fe $3d$, Rh $4d$, and O $2p$ densities of states in the same energy range. Contributions due to orbitals other than those shown are small and can be neglected. The three groups of bands naturally reappear as distinct structures in the DOS. Obviously, the lowest group is dominated by O $2p$ states, whereas the middle group reveals both Fe $3d$ and Rh $4d$ contributions. The former states give rise to a pronounced DOS peak at about -1.2 eV, whereas the strongest DOS peaks due to the latter states are located near -0.3 eV and 0.7 eV. Moreover, the sharp Rh $4d$ e_g peak around 2.2 eV is visible. Apparently, the partial densities of states in figure 7.11 show substantial contributions of the transition metal $3d/4d$ and oxygen $2p$ states below and above -2.0 eV, respectively. The magnitudes of the different admixtures in the energy regions dominated by the respective other states reach up to approximately 30% of the total DOS. In particular, these admixtures are very similar to the results reported for Ca_3CoRhO_6, compare figure 7.4. The p-d hybridizations inherent to the compounds seem to resemble each other. Contributions due to the octahedrally coordinated rhodium sites occur mainly in the interval from -6.7 eV to -5.0 eV, setting up a broad structure. Fe $3d$ admixtures are equally spread over the whole region between -6.7 eV and -2.3 eV. Finally, strong oxygen peaks accompany all the transition metal peaks.
Investigating the partial Rh $4d$ and Fe $3d$ DOS in more detail we identify the crystal field splitting due to the octahedral and trigonal prismatic environments, respectively. For this purpose figure 7.12 depicts the DOS separated into the symmetry components. Here the classification of the rhodium $4d$ DOS in t_{2g} and e_g contributions refers to the local rotated reference frame. As a consequence of the octahedral crystal field one observes an almost perfect splitting into filled t_{2g} and empty e_g states. On the contrary, the trigonal prismatic crystal field at the iron sites gives rise to $3d_{3z^2-r^2}$, $3d_{xy,x^2-y^2}$, and $3d_{xz,yz}$ symmetry components. The latter two states result in sharp peaks directly below and above the Fermi level, whereas the $3d_{3z^2-r^2}$ DOS spans a much wider energy region. As a consequence, the gross features of the $3d$ DOS have not changed in comparison to the Ca_3CoRhO_6 results. However, because iron provides one valence electron less than cobalt, the DOS curves are shifted to higher energies, see figure 7.5. Hence only Fe $3d_{3z^2-r^2}$ states are located in the vicinity of the Fermi energy. The broadening and splitting of the Fe $3d_{3z^2-r^2}$ DOS, and the incomplete crystal field splitting, indicate deviations from an ideal trigonal prismatic coordination, which we mainly ascribe to intrachain metal-metal bonding.
Due to the anisotropy of the Ca_3FeRhO_6 structure the Rh $4d_{3z^2-r^2}$ as well as Fe $3d_{3z^2-r^2}$ orbitals play a central role for understanding the electronic behaviour. They mediate the main overlap along the quasi one-dimensional chains. To analyze the properties of these states more accurately figure 7.13 shows weighted band structures. On the left hand side

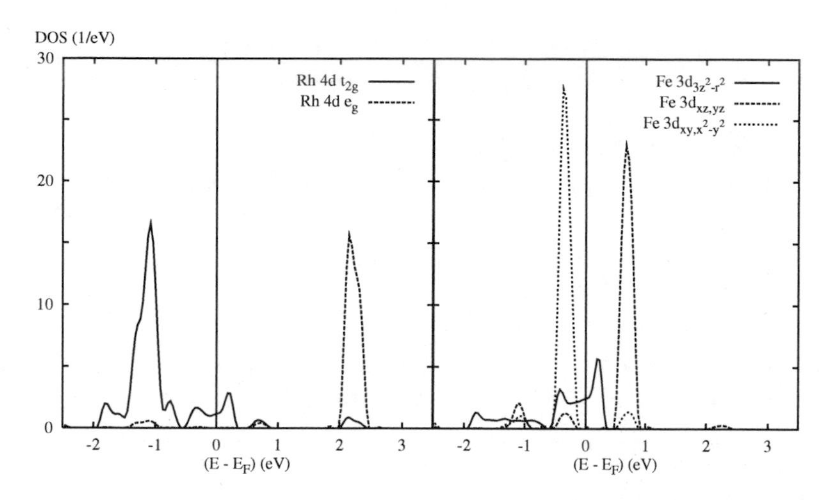

Figure 7.12: *Partial Rh 4d as well as Fe 3d densities of states (DOS) per unit cell of non-magnetic Ca_3CoRhO_6. While the partial DOS at the octahedral rhodium sites separates into 4d t_{2g} and e_g states (with respect to the local rotated reference frame; left), the trigonal prismatic iron sites are represented by the states $3d_{3z^2-r^2}$, $3d_{xz,yz}$, and $3d_{xy,x^2-y^2}$ (right).*

the length of the bars represents the magnitude of the Rh $4d_{3z^2-r^2}$ contributions and the labeling refers to the unrotated Cartesian coordinate system with the z-axis parallel to the quasi one-dimensional Co-Rh chains. The right diagram of figure 7.6 highlights the Fe $3d_{3z^2-r^2}$ contributions. In contrast to the band structure in figure 7.11 we do not use the first Brillouin zone of the primitive trigonal lattice in figure 7.6. Instead, the weighted bands refer to the non-primitive hexagonal representation as given by equation (3.4) and the first Brillouin zone of the hexagonal lattice as illustrated in figure 3.12. This is most useful since the hexagonal symmetry lines reflect specific directions of the crystal lattice. In particular, the hexagonal k_z-axis (line Γ-A) represents the chain direction. Obviously, each of the four $d_{3z^2-r^2}$ bands displays a considerable dispersion along the chains, whereas the dispersion perpendicular is almost negligible. Furthermore, the intrachain dispersions of the $d_{3z^2-r^2}$-type bands are the largest of all the bands shown. Both these observations confirm the structural anisotropy of Ca_3FeRhO_6 since they are indicative of (quasi) one-dimensional electronic states. Compared to the results for Ca_3CoRhO_6, see figure 7.6, the widths of the quasi one-dimensional bands are increased. Furthermore, the dispersions of the other bands are slightly smaller, particularly in the Γ-A direction. Thus the electronic anisotropy seems to be larger in the case of Ca_3FeRhO_6.
Recalling the decomposition of the partial Fe 3d DOS into its symmetry components the bands without bars above and below the Fermy level in figure 7.13 are strongly influenced by the Fe $3d_{xy,x^2-y^2}$ and Fe $3d_{xz,yz}$ states. They reveal an increased dispersion within the hexagonal planes indicative of Fe-O overlap and/or interchain coupling. In contrast, the Rh $4d_{3z^2-r^2}$ and Fe $3d_{3z^2-r^2}$ orbitals mediate overlap along c_{hex}. As the four bands trac-

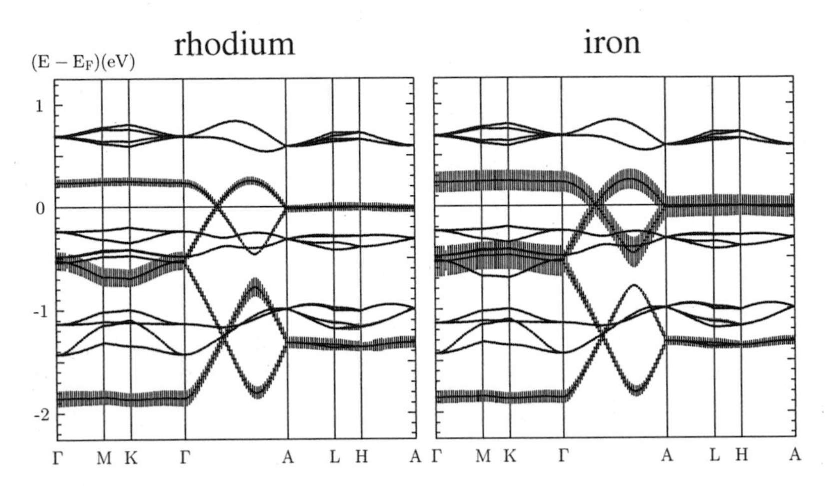

Figure 7.13: *Weighted electronic bands of non-magnetic Ca_3FeRhO_6 shown along selected symmetry lines in the first Brillouin zone of the hexagonal lattice, compare figure 3.6. The bars give the contribution due to the Rh 4$d_{3z^2-r^2}$ (left) and the Fe 3$d_{3z^2-r^2}$ (right) orbital.*

ing back to the $d_{3z^2-r^2}$ orbitals reveal both rhodium and iron contributions we expect a considerable metal-metal bonding along the Fe-Rh chains. The emerging one-dimensional bands cause the discussed deviations from an ideal trigonal crystal field at the iron sites. While the bonding states due to Fe-Rh coupling are rhodium-like, the antibonding branch shows an increased iron character. Quite the opposite has been reported for Ca_3CoRhO_6, where the trigonal prismatic cobalt atoms mainly contributed to the bonding branch. Due to a modified electron count at the trigonal prismatic sites it is not surprising to find the iron-like states at higher energies.
Additional hints at strong metal-metal bonding can be derived from the covalence energy curves depicted in figure 7.14. While metal-oxygen overlap yields bonding contributions mainly below -2.3 eV, the respective antibonding states are located above -2.1 eV. Consistent with the Rh 4d DOS observed in the energy interval from -6.7 eV to -5.0 eV (see figure 7.11) we find negative covalence energies due to overlap between Rh 4d and O 2p orbitals in the same energy region. Contributions tracing back to bonding between Fe 3d and O 2p states predominately are observed between -5.0 eV and -2.3 eV. This observation confirms our expectation from the hybridization effects identified in the study of the DOS. The shape of the Fe-Rh covalence energy curve reflects the results from figure 7.7. Consequently, one identifies the negative bond energy peak around -1.8 eV and the positive peak near 0.2 eV with the bonding and antibonding transition metal $d_{3z^2-r^2}$ states, respectively. Compare the weighted band structures depicted in figure 7.13. Metal-metal covalence energy peaks between -1.5 eV and -0.2 eV are very difficult to interpret since all the different d electrons contribute in this region. Altogether, the covalence energies point to strong in-chain Fe-Rh coupling via the $d_{3z^2-r^2}$ orbitals. In the next step we will

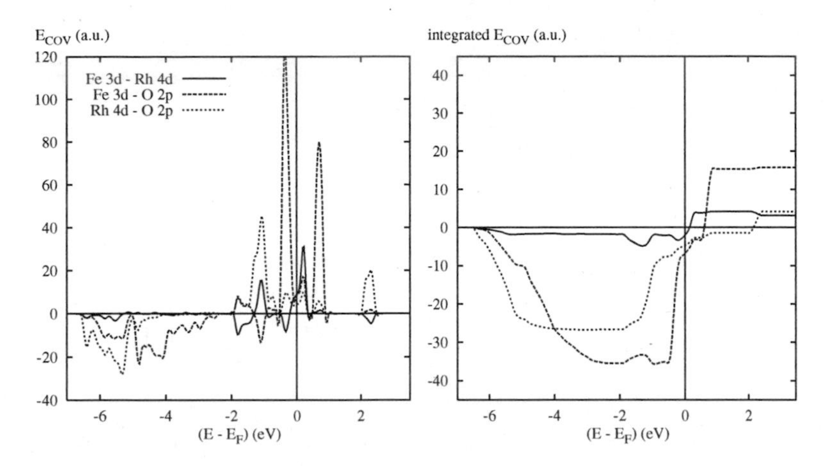

Figure 7.14: *Covalent bond energies of spin degenerate Ca_3FeRhO_6. Negative and positive values point to bonding and antibonding states, respectively. Integrated curves were added.*

investigate the chemical stability of Ca_3FeRhO_6 by means of the integrated covalent bond energies given on the right hand side of figure 7.14. Both Fe-O and Rh-O bonds yield stabilizing net contributions since their covalence bond energies reveal negative values at the Fermi energy. The influences of metal-metal overlap are small. Admittedly, in contrast to our findings in the case of cobalt, Fe-Rh bonding slightly stabilizes the crystal structure. Confirming the previous conclusions, the in-chain bonding and thus the anisotropy of the electronic properties appears to be increased in the case of Ca_3FeRhO_6. Nonetheless, the non-magnetic LDA calculations display similar results for Ca_3CoRhO_6 and Ca_3FeRhO_6. For this reason we analyze the antiferromagnetic in-chain coupling in the latter material in full analogy to our former study of the cobalt compound.

According to the experimental results we model the magnetic coupling in Ca_3FeRhO_6 by an assumed antiferromagnetic configuration. Due to the strong anisotropy of the crystal structure and hence of the electronic states we expect a dominating magnetic exchange coupling along the Fe-Rh chains. Starting with the non-magnetic LDA results it is possible to end with a converged solution with well localized magnetic moments. Per atomic site the following numbers are found: $0.00\,\mu_B$ for rhodium, $3.72\,\mu_B$ for iron, $0.14\,\mu_B$ for oxygen, and $0.01\,\mu_B$ for calcium. Of course, the values apply to both the spin up and spin down sublattice. The total magnetic moments of the sublattices (per unit cell) amount to $\pm 4.59\,\mu_B$. Apart from lifting the spin degeneracy, all technical details are the same as in the non-magnetic case.

The reported magnetic moments for antiferromagnetic Ca_3FeRhO_6 reflect the experimental low and high spin states of the rhodium atoms at the octahedral and the iron atoms at the trigonal prismatic sites, respectively. Figure 7.15 depicts the corresponding partial spin-majority as well as spin-minority densities of states. Consistent with a vanishing mag-

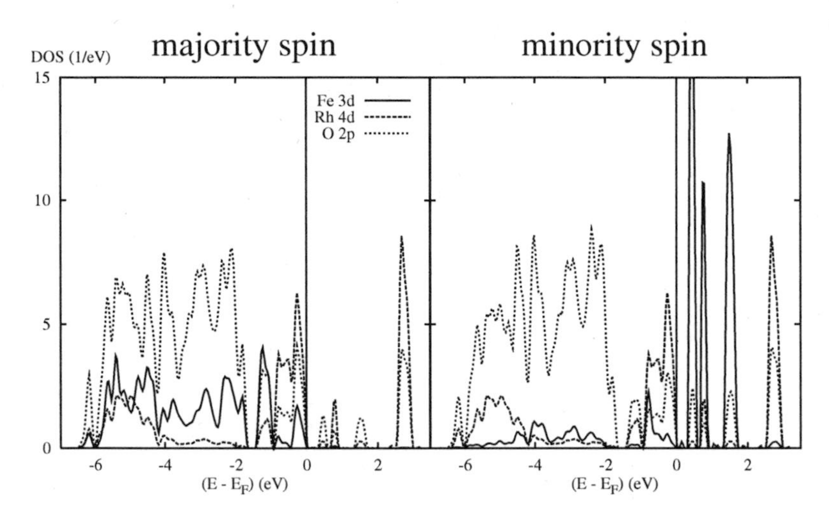

Figure 7.15: *Partial Fe 3d, Rh 4d, and O 2p densities of states (DOS) per unit cell of antiferromagnetic Ca_3FeRhO_6 – separated into spin-majority and spin-minority contributions.*

netic moment, no spin splitting affects the rhodium atoms. In contrast, the spin-majority and spin-minority iron $3d$ densities of states disagree substantially. The missing DOS at the Fermi level clearly characterizes Ca_3FeRhO_6 as an antiferromagnetic insulator. More specifically, one observes an indirect band gap of 0.40 eV. However, this result cannot be compared to experimental data since no resistivity measurements for the iron compound are available in the literature. Because the LDA electronic structure confirms the experimentally determined antiferromagentic in-chain coupling, the calculated insulating state also seems reliable. Comparing the antiferromagnetic configuration of Ca_3FeRhO_6 to the spin degenerate case, the total energy per unit cell decreases by roughly 151 mRyd. As a test, the antiferromagnetic Ca_3FeRhO_6 calculation is complemented (no figures displayed) by a ferromagnetic structure where the magnetism is assumed equivalent to ferromagnetic Ca_3CoRhO_6, see the preceding section. In doing so one obtains a solution with a total energy per unit cell about 3 mRyd higher than in the antiferromagnetic case, which confirms the measured magnetic structure. The ferromagnetic order likewise leads to an insulating energy gap, which amounts to about 0.17 eV. We find the following moments (per atomic site): 0.31 μ_B for rhodium, 3.74 μ_B for iron, 0.14 μ_B for oxygen, and 0.01 μ_B for calcium. Thus the magnetic moment per unit cell is 10.00 μ_B. Assumed ferromagnetic Ca_3FeRhO_6 can be understood analogous with assumed ferromagnetic Ca_3CoRhO_6. In particular, one finds an induced magnetic moment at the rhodium sites because of the strong metal-metal bonding.

Returning to the experimentally predicted antiferromagnetic in-chain coupling we now discuss the findings from figure 7.15 in more detail. While the shape of the spin-minority Fe $3d$ DOS resembles the gross features of the corresponding non-magnetic DOS to some ex-

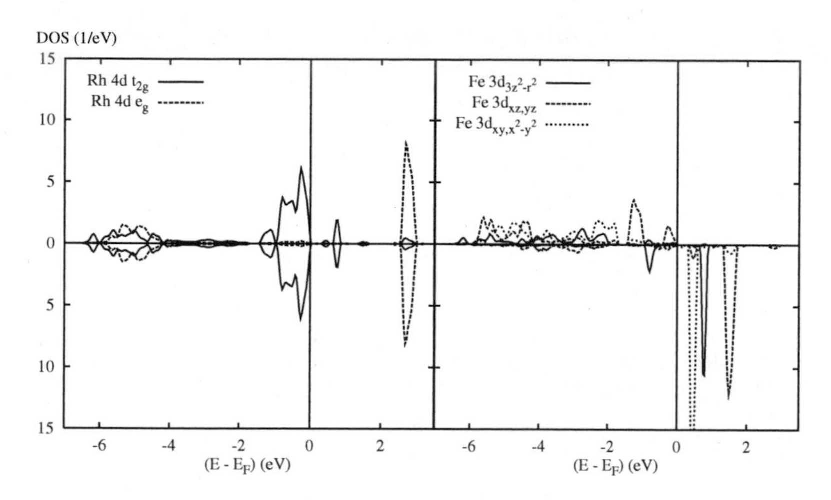

Figure 7.16: *Partial Rh 4d and Fe 3d densities of states (DOS) per unit cell of antiferromagnetic Ca_3FeRhO_6 – separated into the spin-majority (top) and spin-minority (bottom) contributions. While the partial density of states of the octahedral rhodium sites splits up into 4d t_{2g} and e_g states (with respect to the local rotated reference frame; left), the trigonal prismatic iron sites are represented by the states $3d_{3z^2-r^2}$, $3d_{xz,yz}$, and $3d_{xy,x^2-y^2}$ (right).*

tent (figure 7.12), the spin-majority states show strong shifts to lower energies. Throughout the oxygen dominated energy region remarkable Fe 3*d* contributions are recognized. Furthermore, there are strong Rh 4*d* admixtures in the lower oxygen dominated interval from roughly −6.4 eV to −4.2 eV. They are similar for the spin-majority, spin-minority, and non-magnetic DOS. Moreover, we observe contributions from the O 2*p* states above −1.7 eV accompanying all the transition metal peaks in figure 7.15. As discussed for the non-magnetic results, such contributions in the regions where the respective other states dominate trace back to the *p*-*d* hybridization. Especially hybridization between Fe 3*d* and O 2*p* orbitals may account for the high magnetic moment at the oxygen sites. An equivalently strong polarization of the ligand states was found in Ca_3CoRhO_6 and interpreted as extended moment formation.

Aiming at an investigation of the moment formation at the single atomic sites we turn to the partial Rh 4*d* and Fe 3*d* densities of states depicted in figure 7.16. The classification of the Rh 4*d* DOS in t_{2g} and e_g contributions refers to the local rotated reference frame. The figure gives both spin-majority and spin-minority densities of states. Similar to the spin degenerate calculation we recognize rather strong crystal field splittings into 4*d* t_{2g} and 4*d* e_g states at the octahedral rhodium sites. While t_{2g} states appear mainly in the vicinity of the Fermi energy, a pronounced e_g peak is found at roughly 2.7 eV. Due to the hybridization with O 2*p* orbitals additional Rh 4*d* t_{2g} and e_g states occupy a wide area around −5.0 eV. Indeed, the rhodium spin-majority and spin-minority densities of states are equal and consequently no magnetic moment is formed.

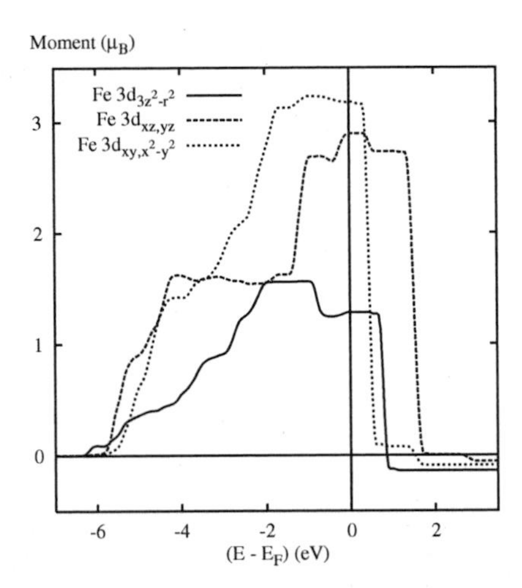

Figure 7.17: *Magnetic Fe 3d moments per unit cell of antiferromagnetic Ca_3FeRhO_6. The notation is the same as used previously for the partial densities of states (see figure 7.16).*

As usual, we distinguish at the iron sites the non-degenerate $3d_{3z^2-r^2}$ as well as the twofold degenerate $3d_{xz,yz}$ and $3d_{xy,x^2-y^2}$ symmetry components. Due to the strong spin splitting at these sites a huge amount of spin-majority Fe $3d$ states is shifted below about $-1.8\,\text{eV}$ giving rise to a very broad DOS shape. Similar to the findings for the cobalt compound this is consistent with an extended moment formation. Spin-minority $3d$ contributions in the energy interval dominated by the oxygen states are much smaller and the respective DOS reveals sharp peaks above the Fermi level. Compared to figure 7.12 the $3d_{xz,yz}$ and $3d_{xy,x^2-y^2}$ peaks appear at higher energies in figure 7.16. Furthermore, the non-magnetic Fe $3d_{3z^2-r^2}$ DOS is characterized by a broader shape than found for the antiferromagnetic configuration. Indeed, a reduced width of the one-dimensional $d_{3z^2-r^2}$ bands is consistent with the antiferromagnetic intrachain ordering. Because spin-majority and spin-minority states are exchanged at neighbouring iron sites (separated by one rhodium atom) the intrachain overlap of the charge carriers decreases. Accordingly, an assumed ferromagnetic structure gives rise to broader $d_{3z^2-r^2}$ bands and the Fe $3d$ DOS resembles qualitatively the findings for the cobalt sites in Ca_3CoRhO_6, see figure 7.9.
In order to analyze the implications of the discussed energetical downshift of the Fe $3d$ spin-majority states, energy and orbital dependent spin magnetizations are displayed in figure 7.17. These spin magnetizations are calculated as the integrated difference of the spin-majority and spin-minority densities of states from figure 7.16. Recall that the Fe $3d$ symmetry components $d_{3z^2-r^2}$, $d_{xz,yz}$, and d_{xy,x^2-y^2} refer to the non-rotated Cartesian coordinate system. Therefore the principal axis of the $d_{3z^2-r^2}$ orbital is aligned parallel to the c_{hex}-axis and consequently points along the Fe-Rh chains. As previously mentioned, the large spin splitting at the iron sites involves all the $3d$ orbitals to some degree. Bear in mind that the $d_{xz,yz}$ and d_{xy,x^2-y^2} symmetry components in figure 7.17 each represent

two atomic orbitals. Summing over the occupied states up to the Fermi energy we obtain a magnetic moment of about $0.65\,\mu_B$ carried by each $d_{3z^2-r^2}$ orbital. Moreover, the $d_{xz,yz}$ moment amounts to approximately $0.75\,\mu_B$ per orbital, whereas the d_{xy,x^2-y^2} moment is $0.80\,\mu_B$ per orbital. In comparison to the findings for Ca_3CoRhO_6 the $d_{3z^2-r^2}$ and $d_{xz,yz}$ moments are slightly larger. However, the main difference between the iron and the cobalt compound is concerned with the d_{xy,x^2-y^2} states. While the latter account for a moment of $0.40\,\mu_B$ for Ca_3CoRhO_6, we observe an increased value for Ca_3FeRhO_6. Thus the distribution of the magnetic moment between the five Co/Fe $3d$ orbitals is more isotropic in the case of iron.
Comparing the energy and orbital dependent spin magnetizations in figures 7.10 and 7.17 we can identify the influence of the reduced electron count at the trigonal prismatic sites. As iron provides one electron less than cobalt, all curves shift to higher energies in figure 7.17. The shapes of corresponding magnetizations are very similar, particularly near the Fermi level. By means of the energetical upshift the right edge of the d_{xy,x^2-y^2} curve shifts across the Fermi energy therefore giving rise to a doubling of the magnetic moment. This increase can already be identified in the partial DOS, see figures 7.9 and 7.16. While the most distinct spin-minority Co $3d_{xy,x^2-y^2}$ peak is located right at the Fermi level, the corresponding Fe $3d_{xy,x^2-y^2}$ states shift to higher energies. Since less spin-minority d_{xy,x^2-y^2} states are occupied the resulting magnetic moment increases. Similar to the findings for Ca_3CoRhO_6, figure 7.17 is indicative of extended moment formation at the trigonal prismatic sites: the magnetic moment starts to accumulate at about $-6.0\,\mathrm{eV}$. In comparison to the cobalt compound the accumulation of moment in the energy region dominated by the O $2p$ states is even enhanced. Due to the strong p-d hybridization mentioned in the discussion of the spin degenerate Ca_3FeRhO_6 calculation the O $2p$ orbitals can be easily polarized, which yields the considerable magnetic moment of $0.14\,\mu_B$ at the oxygen sites. A strong O $2p$ polarization resembles well the Ca_3CoRhO_6 results. The same is true for the fact that the magnetism of Ca_3FeRhO_6 seems to be triggered mainly by the atoms at trigonal prismatic sites. Replacing cobalt (d^6) by iron (d^5) affects the magnetic properties since the electron count is changed. As the O $2p$ orbitals participate in the local magnetic moments, but not in the spin exchange coupling between the trigonal prisms, we are confronted with extended moment formation. Coupling between the extended (but localized) moments is mediated by superexchange interaction via $d_{3z^2-r^2}$ orbitals. Although the octahedral atoms do not contribute to the magnetic moment they hence are most important for the exchange coupling.
Since the electronic configuration of the trigonal prismatic atoms seriously influences the realized magnetic structure we now analyze the particular oxidation states calculated for Ca_3FeRhO_6. The formal [Ar]$3d^64s^2$ set-up of iron yields a reduced electron count compared to Ca_3CoRhO_6, which is characterized by the formal [Ar]$4d^75s^2$ set-up of its cobalt atoms. In tabel 7.3 calculated LDA valence charges are given for both the assumed ferromagnetic configuration of Ca_3CoRhO_6 and the assumed antiferromagnetic configuration of Ca_3FeRhO_6. When dealing with valences we bear in mind the limitations of assigning electronic charge to particular atomic sites. As the radii of the transition metal spheres are similar, the comparison is more reliable and we find the oxidation states $Rh^{2.08+}$ and $Fe^{1.91+}$ for Ca_3FeRhO_6. Remember the corresponding findings for the cobalt compound: $Rh^{2.24+}$ and $Co^{1.80+}$. The transfer of charge from the trigonal prismatic to the octahedral

sites is thus somewhat increased in Ca_3FeRhO_6. In both materials the calculated valences differ significantly from the ideal ionic configuration $Ca_3^{2+}(Co/Fe)^{3+}Rh^{3+}O_6^{2-}$ because of strong covalent bonding. Assuming the oxidation state 3+ for all transition metal atoms in Ca_3FeRhO_6 yields five electrons at the iron and six electrons at the rhodium sites. In the latter case the octahedral coordination leads to occupied t_{2g} orbitals and hence to a $S = 0$ low spin state. For the trigonal prismatic sites we expect one electron in each of the five Fe 3*d* orbitals, which implies a $S = 5/2$ high spin state. The LDA calculation yields a magnetic moment of $3.72\,\mu_B$ per iron site. Contradicting this result, the experimentally determined saturation magnetization is $3.74\,\mu_B$ per formula unit and hence significantly smaller than expected. But the LDA calculation yields magnetic moments not only at the iron sites but also at the oxygen sites due to Fe-O hybridization. We must therefore also take into account the six oxygen moments per formula unit, each amounting to $0.14\,\mu_B$. Because of substantial deviations in the stoichiometry of the probed samples, referring to S. Niitaka *et al.* [154], the measured saturation magnetization might be much too small. In contrast to Ca_3CoRhO_6 no magnetic moment is found at the octahedral sites of the iron compound. In the cobalt compound the spin polarization at these sites appears to be induced by the trigonal prismatic atoms due to strong in-chain metal-metal bonding. Apparently, the antiferromagnetic arrangement of the iron moments in Ca_3FeRhO_6 does not allow for this mechanism.

Chapter 8

Summary and Outlook

In the present thesis 'first principles' calculations have been used to study the interplay of the structural and electronic properties in transition metal compounds. In particular, the influence of electron-lattice coupling in vanadium and titanium oxides, structural deviations in perovskite-related materials, and the spin moments in magnetic chain compounds have been analyzed. Combining insight into local atomic coordination, chemical bonding, and electronic structure has paved way for understanding most exciting material properties. In all the investigated compounds we observed strong hybridization effects between transition metal $3d/4d$ and oxygen $2p$ states. Hybridization combined with crystal field splitting at the transition metal sites and effects of covalent metal-metal bonding leads to unusual electronic structures, giving rise to interesting technological applications.
In order to investigate the metal-insulator transitions of the prototypical vanadium oxides VO_2 and V_2O_3, we have studied the Magnéli phases V_nO_{2n-1}. Analyzing the electronic states involved in the phase transitions has allowed us to discuss the effects of electronic correlations and electron-lattice interaction when going from VO_2 to V_2O_3. The Magnéli phases are a homologous series of compounds with crystal structures comprising typical dioxide and sesquioxide-like regions. We have developed a unifying description of the different crystal structures including the rutile structure of VO_2 and the corundum structure of V_2O_3. This representation is based on a regular three-dimensional network of oxygen octahedra partially filled by metal atoms – hence giving rise to metal chains of length n. Using the comprehensive picture of the crystal structures we have been able to group the electronic bands in states behaving either as VO_2 or as V_2O_3.
While the overlap of vanadium and oxygen places the V $3d$ t_{2g} states in the vicinity of the Fermi energy, the detailed electronic structure and thus the metal-insulator transitions in the Magnéli phases have been found to be essentially influenced by the local metal-metal coordinations. The phase transitions arise as a result of both electron-lattice interaction within the dioxide-like and electronic correlations within the sesquioxide-like regions of the crystal. Dioxide-like vanadium atoms show the characteristical features of the embedded Peierls instability responsible for the metal-insulator transition of VO_2. A combination of dimerization and antiferroelectric-like displacements of metal atoms by means of strong electron-lattice interaction causes a splitting of the $d_{x^2-y^2} = d_{\parallel}$ states and an energetical upshift of the d_{yz} as well as d_{xz} states. Sites related to the sesquioxide are characterized

by their strongly reduced metal-metal overlap. Therefore the $d_{x^2-y^2}$ orbitals localize and their partial densities of states resemble those of the d_{xz} orbitals. Due to the localization electronic correlations might play an important role. The electronic structures and phase transitions of the vanadium Magnéli series are a crucial test case for theories aiming at a correct description of both vanadium dioxide and sesquioxide.
We have succeeded in transferring findings for the V_2O_3-like metal atoms of the Magnéli phases to vanadium sesquioxide itself. In particular, metal-metal bonding along the c_{hex}-axis of V_2O_3 causes the major effect on the shape of the a_{1g} density of states. However, the V-V interactions perpendicular to the c_{hex}-axis are also large since they yield strong splittings of the electronic states. Because of reduced metal-metal overlap across the octahedral edges the V $d_{x^2-y^2}$ orbitals are more localized. Moreover, they undergo further localization due to (local) structural modifications typical for the insulating phase. This band narrowing paves way for an increased influence of the electronic correlations, which are regarded to be most important for the phase transitions of V_2O_3. None of the effects could be found in previous studies of the sesquioxide due to its high symmetry. Nonetheless, they are essential to properly understand the phase transitions of V_2O_3. Comparing the electronic structures of vanadium and titanium sesquioxide has permitted us to analyze the complex V $3d$ t_{2g} density of states. Because of V-V overlap along the c_{hex}-axis the quasi one-dimensional a_{1g} states split into bonding and antibonding branches. However, the wide bonding branch is strongly affected by a_{1g}-e_g^π coupling. A reasonable treatment of the a_{1g} states consequently requires a model going beyond a simple bonding-antibonding split one-dimensional band.
Due to very similar crystal structures it has been possible to transfer the knowledge of the phase transitions in the vanadium Magnéli systems to the titanium oxides. In particular, the metal-insulator transition of Ti_4O_7 resembles almost all the characteristics reported for V_4O_7. Half the titanium chains are subject to dimerization and antiferroelectric-like displacements. The other chains are characterized by a strong localization of the $d_{x^2-y^2}$ orbitals due to significantly reduced intrachain metal-metal overlap in the low-temperature phase. Charge transfer from the localizing to the dimerizing chains eventually allows for an embedded Peierls instability on the latter. The metal-insulator transition hence arises from a complex interplay of charge order, orbital order, and electron localization.
The electronic properties of the perovskite-related ruthenates $ACu_3Ru_4O_{12}$ (A=Na, Ca, Sr, La, Nd) are governed by strong covalent bonding between transition metal $3d/4d$ and oxygen $2p$ orbitals. Via the bond lengths, Ru-O interaction affects the size of the oxygen octahedra and hence the (cubic) lattice constant. Octahedral tilting traces back to Cu-O bonding as it hardly influences the Ru-O bonds, whereas the Cu-O bond lengths undergo serious changes, eventually leading to a square-planar coordination of copper. The A-O distances are exceptionally small because the response of the A-O bonds to the tilting is small and the structure does not offer any other degree of freedom to optimize the A-O bond lengths. In conclusion, octahedral tilting is a consequence of the covalent bonding between different atomic species resulting in optimal bond lengths. As this is a very general principle it applies to a large variety of compounds and is therefore most useful for tailoring materials with specific properties.
The studied magnetic chain compounds Ca_3CoRhO_6 and Ca_3FeRhO_6 have shown strong effects of the local transition metal coordination on the electronic and magnetic structure.

While the intrachain ordering of the cobalt sites in Ca_3CoRhO_6 is ferromagnetic, the iron sites of Ca_3FeRhO_6 order antiferromagnetically. Due to strong Rh-O bonding within the oxygen octahedra only small spin splittings are found at the rhodium sites of Ca_3CoRhO_6. The rhodium atoms of Ca_3FeRhO_6 carry no magnetic moment. Because the crystal field splitting at the trigonal prismatic Co/Fe sites is reduced these atoms exhibit a high spin state. Hybridization effects between Co/Fe $3d$ and O $2p$ states cause large magnetic moments at the oxygen sites, giving rise to extended moment formation. A distinct overlap of the transition metal $d_{3z^2-r^2}$ states forms the basis of superexchange coupling between the extended moments – mediated by low spin Rh $d_{3z^2-r^2}$ orbitals. Neither the extended moment formation nor the mechanism of the magnetic coupling changes with the realized magnetic order. In general, the intrachain magnetism of Ca_3CoRhO_6 and Ca_3FeRhO_6 is well described in terms of electronic band structure calculations. Comparing the exchange interactions in these materials appears to be most promising for better understanding the complex phase diagrams of a large class of magnetic chain compounds.

Bibliography

[1] R. O. Jones and O. Gunnarsson, *The density functional formalism, its applications and prospects*, Reviews of Modern Physics **61**, 689 (1989).

[2] R. G. Parr and W. Yang, *Density-Functional Theory of Atoms and Molecules* (Oxford University Press, New York, 1989).

[3] R. M. Dreizler and E. K. U. Gross, *Density Functional Theory – An Approach to the Quantum Many-Body Problem* (Springer, Berlin, 1990).

[4] E. K. U. Gross and R. M. Dreizler (Eds.), *Density Functional Theory* (Plenum Press, New York, 1995).

[5] H. Eschrig, *The Fundamentals of Density Functional Theory* (Teubner, Stuttgart, 1996).

[6] M. Springborg (Ed.), *Density-Functional Methods in Chemistry and Materials Science* (Wiley, Chichester, 1997).

[7] M. Springborg, *Methods of Electronic-Structure Calculations – From Molecules to Solids* (Wiley, Chichester, 2000).

[8] P. L. Taylor and O. Heinonen, *A Quantum Approach to Condensed Matter Physics* (Cambridge University Press, Cambridge 2002).

[9] M. Born and R. Oppenheimer, *Zur Quantentheorie der Molekeln*, Annalen der Physik **84**, 457 (1927).

[10] P. Hohenberg and W. Kohn, *Inhomogeneous Electron Gas*, Physical Review **136**, B864 (1964).

[11] M. Levy, *Electron densities in search of Hamiltonians*, Physical Review A **26**, 1200 (1982).

[12] W. Nolting, *Grundkurs theoretische Physik*, Volume 5/2 (Springer, Berlin, 2002).

[13] W. Kohn and L. J. Sham, *Self-Consistent Equations Including Exchange and Correlation Effects*, Physical Review **140**, A1133 (1965).

[14] J. F. Janak, *Proof that $\partial E/\partial n_i = \epsilon_i$ in density-functional theory*, Physical Review B **18**, 7165 (1978).

[15] C.-O. Almbladh and U. von Barth, *Exact results for the charge and spin densities, exchange-correlation potentials, and density-functional eigenvalues*, Physical Review B **31**, 3231 (1985).

[16] U. von Barth and L. Hedin, *A local exchange-correlation potential for the spin polarized case: I*, Journal of Physics C: Solid State Physics **7**, 1629 (1972).

[17] A. K. Rajagopal and J. Callaway, *Inhomogeneous Electron Gas*, Physical Review B **7**, 1912 (1973).

[18] J. Kübler and V. Eyert, *Electronic structure calculations*, in K. H. J. Buschow (Ed.), *Electronic and Magnetic Properties of Metals and Ceramics* (Wiley-VCH, Weinheim, 1992).

[19] J. M. MacLaren, D. P. Clougherty, M. E. McHenry, and M. M. Donovan, *Parameterised local spin density exchange-correlation energies and potentials for electronic structure calculations – I. Zero temperature formalism*, Computer Physics Communications **66**, 383 (1991).

[20] S. H. Vosko, L. Wilk, and M. Nusair, *Accurate spin-dependent electron liquid correlation energies for local spin density calculations: a critical analysis*, Canadian Journal of Physics **58**, 1200 (1980).

[21] J. P. Perdew and A. Zunger, *Self-interaction correction to density-functional approximations for many-electron systems*, Physical Review B **23**, 5048 (1981).

[22] J. P. Perdew and M. Levy, *Physical Content of the Exact Kohn-Sham Orbital Energies: Band Gaps and Derivative Discontinuities*, Physical Review Letters **20**, 1884 (1983).

[23] L. J. Sham and M. Schlüter, *Density-Functional Theory of the Energy Gap*, Physical Review Letters **20**, 1888 (1983).

[24] V. Eyert, *Basic Notions and Applications of the Augmented Spherical Wave Method*, International Journal of Quantum Chemistry **77**, 1007 (2000).

[25] J. C. Slater, *Wave Functions in a Periodic Potential*, Physical Review **51**, 846 (1937).

[26] O. K. Andersen, *Linear methods in band theory*, Physical Review B **12**, 3060 (1975).

[27] A. R. Williams, J. Kübler, and C. D. Gelatt Jr., *Cohesive properties of metallic compounds: Augmented-spherical-wave calculations*, Physical Review B **19**, 6094 (1979).

[28] V. Eyert, *Entwicklung und Implementation eines Full-Potential-ASW-Verfahrens*, PhD-thesis (Technische Hochschule Darmstadt, 1991).

[29] R. Hoffmann, *A Chemist's View of Bonding in Extended Structures* (VCH, New York, 1988).

[30] N. Börnsen, B. Meyer, O. Grotheer and M. Fähnle, *E_{cov} – a new tool for the analysis of electronic structure data in a chemical language*, Journal of Physics: Condensed Matter **11**, L287 (1999).

[31] R. Dronskowski and P. E. Blöchl, *Crystal orbital Hamilton populations (COHP): energy-resolved visualization of chemical bonding in solids based on density-functional calculations*, Journal of Physical Chemistry **97**, 8617 (1993).

[32] J. B. Goodenough, *Metallic Oxides*, Progress in Solid State Chemistry **5**, 145 (1971).

[33] W. Brückner, H. Oppermann, W. Reichelt, J. I. Terukow, F. A. Tschudnowski, and E. Wolf, *Vanadiumoxide – Darstellung, Eigenschaften, Anwendung* (Akademie-Verlag, Berlin, 1983).

[34] M. Imada, A. Fujimori, and Y. Tokura, *Metal-insulator transitions*, Reviews of Modern Physics **70**, 1039 (1998).

[35] D. B. McWhan, M. Marezio, J. P. Remeika, and P. D. Dernier, *X-ray diffraction study of metallic VO_2*, Physical Review B **10**, 490 (1974).

[36] P. I. Sorantin and K. Schwarz, *Chemical Bonding in Rutile-Type Compounds*, Inorganic Chemistry **31**, 567 (1992).

[37] M. Marezio, D. B. McWhan, J. P. Remeika, and P. D. Dernier, *Structural Aspects of the Metal-Insulator Transitions in Cr-Doped VO_2*, Physical Review B **5**, 2541 (1972).

[38] J. P. Pouget, H. Launois, J. P. D'Haenens, P. Merenda, and T. M. Rice, *Electron Localization Induced by Uniaxial Stress in Pure VO_2*, Physical Review Letters **35**, 873 (1975).

[39] J. P. Pouget and H. Launois, *Metal-Insulator Phase Transition in VO_2*, Journal de Physique **37** (C4), 49 (1976).

[40] F. J. Morin, *Oxides Which Show a Metal-to-Insulator Transition at the Neel Temperature*, Physical Review Letters **3**, 34 (1959).

[41] J. M. Longo and P. Kierkegaard, *A Refinement of Structure of VO_2*, Acta Chemica Scandinavica **24**, 420 (1970).

[42] V. Eyert, R. Horny, K.-H. Höck, and S. Horn, *Embedded Peierls instability and the electronic structure of MoO_2*, Journal of Physics: Condensed Matter **12**, 4923 (2000).

[43] V. Eyert, *The metal-insulator transition of NbO_2: An embedded Peierls instability*, Europhysics Letters **58**, 851 (2002).

[44] V. Eyert, *The metal-insulator transitions of VO_2: A band theoretical approach*, Annalen der Physik **11**, 648 (2002).

[45] R. E. Peierls, *Quantum Theory of Solids* (Clarendon Press, Oxford, 1955).

[46] G. Grüner, *The dynamics of charge-density waves*, Reviews of Modern Physics **60**, 1129 (1988).

[47] N. F. Mott, *Metal-Insulator Transition*, Reviews of Modern Physics **40**, 677 (1968).

[48] F. Gebhard, *The Mott Metal-Insulator Transition – Models and Methods* (Springer, Berlin, 1997).

[49] D. Maurer, A. Leue, R. Heichele, and V. Müller, *Elastic behavior near the metal-insulator transition of* VO_2, Physical Review B **60**, 13249 (1999).

[50] H. W. Verleur, A. S. Barker Jr., and C. N. Berglund, *Optical Properties of* VO_2 *between 0.25 and 5 eV*, Physical Review **172**, 788 (1968).

[51] E. Goering, M. Schramme, O. Müller, R. Barth, H. Paulin, M. Klemm, M. L. den Boer, and S. Horn, *LEED and photoemission study of the stability of* VO_2 *surfaces*, Physical Review B **55**, 4225 (1997).

[52] O. Müller, E. Goering, J.-P. Urbach, T. Weber, H. Paulin, M. Klemm, M. L. den Boer, and S. Horn, *Metal-Insulator Transition of* VO_2*: A XANES Investigation of the O K Edge of* VO_2, Journal de Physique IV **7** (C2), 533 (1997).

[53] S. Shin, S. Suga, M. Taniguchi, M. Fujisawa, H. Kanzaki, A. Fujimori, H. Daimon, Y. Ueda, K. Kosuge, and S. Kachi *Vacuum-ultraviolet reflectance and photoemission study of the metal-insulator phase transitions in* VO_2, V_6O_{13}, *and* V_2O_3, Physical Review B **41**, 4993 (1990).

[54] A. Zylbersztejn and N. F. Mott, *Metal-insulator transition in vanadium dioxide*, Physical Review B **11**, 4383 (1975).

[55] R. M. Wentzcovitch, W. W. Schulz, and P. B. Allen, VO_2*: Peierls or Mott-Hubbard? A View from Band Theory*, Physical Review Letters **72**, 3389 (1994); Comment by T. M. Rice, H. Launois, and J. P. Pouget, Physical Review Letters **73**, 3042 (1994); Reply by R. M. Wentzcovitch, W. W. Schulz, and P. B. Allen, Physical Review Letters **73**, 3043 (1994).

[56] W.-D. Yang, *Instability of Local Oxygen Structure and Metal-Insulator Transition in* VO_2 *and* Ti_2O_3, PhD-thesis (Universität Augsburg, 1999).

[57] M. S. Laad, L. Craco, and E. Müller-Hartmann, *On the First-Order Insulator-Metal Transition in the Rutile-based* VO_2, arXiv:cond-mat/0305081.

[58] A. Liebsch and H. Ishida, *Role of Coulomb Correlations in the Metal Insulator Transition of* VO_2, arXiv:cond-mat/0310216.

[59] V. Eyert, *Octahedral Deformations and Metal-Insulator Transition in Transition Metal Calcogenides*, Habilitation thesis (Universität Augsburg, 1998).

[60] C. J. Bradley and A. P. Cracknell, *The Mathematical Theory of Symmetry in Solids* (Clarendon Press, Oxford, 1972).

[61] D. B. McWhan, J. P. Remeika, T. M. Rice, W. F. Brinkman, J. P. Maita, and A. Menth, *Electronic Specific Heat of Metallic Ti-Doped* V_2O_3, Physical Review Letters **27**, 941 (1971).

[62] D. B. McWhan, A. Menth, J. P. Remeika, W. F. Brinkman, and T. M. Rice, *Metal-Insulator Transitions in Pure and Doped* V_2O_3, Physical Review B **7**, 1920 (1973).

[63] H. Kuwamoto, J. M. Honig, and J. Appel, *Electrical properties of the* $(V_{1-x}Cr_x)_2O_3$ *system*, Physical Review B **22**, 2626 (1980).

[64] P. D. Dernier, *The Crystal Structure of* V_2O_3 *and* $(V_{0.962}Cr_{0.038})_2O_3$ *Near the Metal-Insulator Transition*, Journal of Physics and Chemistry of Solids **31**, 2569 (1970).

[65] P. D. Dernier and M. Marezio, *Crystal Structure of the Low-Temperature Antiferromagnetic Phase of* V_2O_3, Physical Review B **2**, 3771 (1970).

[66] N. W. Ashcroft and N. D. Mermin, *Solid State Physics* (Saunders College, Fort Worth, 1976).

[67] P. Pfalzer, J. Will, A. Nateprov Jr., M. Klemm, V. Eyert, S. Horn, A. I. Frenkel, S. Calvin, and M. L. den Boer, *Local symmetry breaking in paramagnetic insulating* $(Al, V)_2O_3$, Physical Review B **66**, 085119 (2002).

[68] C. Castellani, C. R. Natoli, and J. Ranninger, *Magnetic structure of* V_2O_3 *in the insulating phase*, Physical Review B **78**, 4945 (1978); *Insulating phase of* V_2O_3*: An attempt at a realistic calculation*, Physical Review B **78**, 4967 (1978); *Metal-insulator transition in pure and Cr-doped* V_2O_3, Physical Review B **78**, 5001 (1978).

[69] S. Shin, Y. Tezuka, T. Kinoshita, T. Ishii, T. Kashiwakura, M. Takahashi, and Y. Suda, *Photoemission Study of the Spectral Function of* V_2O_3 *in Relation to the Recent Quantum Monte Carlo Study*, Journal of the Physical Society of Japan **64**, 1230 (1995).

[70] K. E. Smith and V. E. Henrich, *Photoemission study of composition- and temperature-induced metal-insulator transitions in Cr-doped* V_2O_3, Physical Review B **50**, 1382 (1994).

[71] O. Müller, J. P. Urbach, E. Goering, T. Weber, R. Barth, H. Schuler, M. Klemm, S. Horn, and M. L. den Boer, *Spectroscopy of metallic and insulating* V_2O_3, Physical Review B **23**, 15056 (1997).

[72] L. F. Mattheiss, *Band properties of metallic corundum-phase* V_2O_3, Journal of Physics: Condensed Matter **6**, 6477 (1994).

[73] J.-H. Park, L.-H. Tjeng, A. Tanaka, J. W. Allen, C. T. Chen, P. Metcalf, J. M. Honig, F. M. F. de Groot, and G. A. Sawatzky, *Spin and orbital occupation and phase transitions in* V_2O_3, Physical Review B **61**, 11506 (2000).

[74] S. Yu. Ezhov, V. I. Anisimov, D. I. Khomskii, and G. A. Sawatzky, *Orbital Occupation, Local Spin, and Exchange Interactions in V_2O_3*, Physical Review Letters **83**, 4136 (1999).

[75] F. Mila, R. Shiina, F.-C. Zhang, A. Joshi, M. Ma, V. Anisimov, and T. M. Rice, *Orbitally Degenerate Spin-1 Model for Insulating V_2O_3*, Physical Review Letters **85**, 1714 (2000).

[76] I. S. Elfimov, T. Saha-Dasgupta, and M. A. Korotin, *Role of x-axis pairs in V_2O_3 from the band-structure point of view*, arXiv:cond-mat/0303404.

[77] W. Bao, C. Broholm, G. Aeppli, P. Dai, J. M. Honig, and P. Metcalf, *Dramatic Switching of Magnetic Exchange in a Classic Transition Metal Oxide: Evidence for Orbital Ordering*, Physical Review Letters **78**, 507 (1997).

[78] K. Held, G. Keller, V. Eyert, D. Vollhardt, and V. I. Anisimov, *Mott-Hubbard Metal-Insulator Transition in Paramagnetic V_2O_3: An LDA+DMFT (QMC) Study*, Physical Review Letters **86**, 5345 (2001).

[79] K. Held, I. A. Nekrasov, G. Keller, V. Eyert, N. Blümer, A. K. McMahan, R. T. Scalettar, T. Pruschke, V. I. Anisimov, D. Vollhardt, *The LDA+DMFT Approach to Materials with Strong Electronic Correlations*, in J. Grotendorst, D. Marx and A. Muramatsu (Eds.), *Quantum Simulations of Complex Many-Body Systems: From Theory to Algorithms* (John von Neumann Institute for Computing, Jülich, 2002).

[80] S.-K. Mo, J. D. Denlinger, H.-D. Kim, J.-H. Park, J. W. Allen, A. Sekiyama, A. Yamasaki, K. Kadono, S. Suga, Y. Saitoh, T. Muro, P. Metcalf, G. Keller, K. Held, V. Eyert, V. I. Anisimov, and D. Vollhardt, *Prominent Quasiparticle Peak in the Photoemission Spectrum of the Metallic Phase of V_2O_3*, Physical Review Letters **90**, 186403 (2003).

[81] M. S. Laad, L. Craco, and E. Müller-Hartmann, *Orbital Switching and the First-Order Insulator-Metal Transition in Paramagnetic V_2O_3*, arXiv:condmat/0211210.

[82] A. Magnéli, *The Crystal Structures of Mo_9O_{26} (β'-Molybdenum Oxide) and Mo_8O_{23} (β-Molybdenum Oxide)*, Acta Chemica Scandinavica **2**, 501 (1948).

[83] S. Andersson and L. Jahnberg, *Crystal structure studies on the homologous series Ti_nO_{2n-1}, V_nO_{2n-1} and $Ti_{n-2}Cr_2O_{2n-1}$*, Arkiv för Kemi **21**, 413 (1963).

[84] S. Kachi, K. Kosuge, and H. Okinaka, *Metal-Insulator Transition in V_nO_{2n-1}*, Journal of Solid State Chemistry **6**, 258 (1973).

[85] G. D. Khattak, P. H. Keesom, and S. P. Faile, *Specific heats of the vanadium Magnéli phases V_nO_{2n-1} between 0.4 and 50 K*, Physical Review B **18**, 6181 (1978).

[86] S. Nagata, P. H. Keesom, and S. P. Faile, *Susceptibilities of the vanadium Magnéli phases V_nO_{2n-1} at low temperature*, Physical Review B **20**, 2886 (1979).

[87] A. C. Gossard, J. P. Remeika, T. M. Rice, and H. Yasuoka, *Microscopic magnetic properties of metallic and insulating V_4O_7 and V_7O_{13}*, Physical Review B **9**, 1230 (1974).

[88] M. Marezio, D. B. McWhan, P. D. Dernier, and J. P. Remeika, *Charge Localization at Metal-Insulator Transitions in Ti_4O_7 and V_4O_7*, Physical Review Letters **28**, 1390 (1972).

[89] B. F. Griffing, S. A. Shivashankar, S. Nagata, S. P. Faile, and J. M. Honig, *Physical properties of V_7O_{13} single crystals*, Physical Review B **25**, 1703 (1982).

[90] B. F. Griffing, S. A. Shivashankar, S. P. Faile, and J. M. Honig, *Metal-insulator transition in V_4O_7: Specific-heat measurements and interpretation*, Physical Review B **31**, 8143 (1985).

[91] A. C. Gossard, F. J. Di Salvo, L. C. Erich, J. P. Remeika, H. Yasuoka, K. Kosuge, and S. Kachi, *Microscopic magnetic properties of vanadium oxides. II. V_3O_5, V_5O_9, V_6O_{11}, and V_6O_{13}*, Physical Review B **10**, 4178 (1974).

[92] P. C. Canfield, J. D. Thompson, and G. Gruner, *Unifying trends found for the V_NO_{2N-1} series by the application of hydrostatic pressure*, Physical Review B **41**, 4850 (1990).

[93] H. Horiuchi, M. Tokonami, N. Morimoto, and K. Nagasawa, *The Crystal Structure of V_4O_7*, Acta Crystallographica **B28**, 1404 (1972).

[94] M. Marezio, D. B. McWhan, P. D. Dernier, J. P. Remeika, *Structural Aspects of the Metal-Insulator Transition in V_4O_7*, Journal of Solid State Chemistry **6**, 419 (1973).

[95] J.-L. Hodeau and M. Marezio, *The Crystal Structure of V_4O_7 at* 120° *K*, Journal of Solid State Chemistry **23**, 253 (1978).

[96] P. C. Canfield, *Exotic Ground States of V_NO_{2N-1} Vanadium Oxides*, PhD-thesis (University of California, Los Angeles, 1990).

[97] Y. Le Page und P. Strobel, *Structural Chemistry of the Magnéli Phases Ti_nO_{2n-1} ($4 \leq n \leq 9$). - I. Cell and Structure Comparisons*, Journal of Solid State Chemistry, **43**, 314 (1982).

[98] H. Horiuchi, N. Morimoto, and M. Tokonami, *Crystal Structures of V_n-O_{2n-1}* ($2 \leq n \leq 7$), Journal of Solid State Chemistry **17**, 407 (1976).

[99] T. Hahn (Ed.), *International Tables for Crystallography – Volume A – Space Group Symmetry* (Kluwer Academic, Dordrecht, 1989).

[100] S. Åsbrink, *The Crystal Structure of and Valency Distribution in the Low-Temperature Modification of V_3O_5. The Decisive Importance of a Few Very Weak Reflexions in a Crystal-Structure Determination*, Acta Crystallographica **B36**, 1332 (1980).

[101] S. Hong und S. Åsbrink, *The Structure of the High-Temperature Modification of V_3O_5 at 458 K*, Acta Crystallographica **B38**, 713 (1982).

[102] V. Eyert and K.-H. Höck, *Electronic Structure of V_2O_5: Role of octahedral deformations*, Physical Review B **57**, 12727 (1998).

[103] U. Schwingenschlögl, V. Eyert, and U. Eckern, *From VO_2 to V_2O_3: The metal-insulator transition of the Magnéli Phase V_6O_{11}*, Europhysics Letters **61**, 361 (2003).

[104] U. Schwingenschlögl, V. Eyert, and U. Eckern, *The Metal-Insulator Transition of the Magnéli Phase V_4O_7: Implications for V_2O_3*, Europhysics Letters **64**, 682 (2003).

[105] Y. Le Page, P. Bordet, and M. Marezio, *Valence Ordering in V_5O_9 Below 120 K*, Journal of Solid State Chemistry **92**, 380 (1991).

[106] M. Marezio, P. D. Dernier, D. B. McWhan, and S. Kachi, *Structural Aspects of the Metal-Insulator Transition in V_5O_9*, Journal of Solid State Chemistry **11**, 301 (1974).

[107] A. D. Inglis, Y. Le Page, P. Strobel, and C. M. Hurd, *Electrical conductance of crystalline Ti_nO_{2n-1} for $n = 4 - 9$*, Journal of Physics C: Solid State Physics **16**, 317 (1983).

[108] V. Eyert, U. Schwingenschlögl, and U. Eckern, *Charge order, orbital order, and electron localization in the Magnéli phase Ti_4O_7*, Chemical Physics Letters, submitted.

[109] R. F. Bartholomew and D. R. Frankl, *Electrical Properties of Some Titanium Oxides*, Physical Review **187**, 828 (1969).

[110] C. Schlenker, S. Lakkis, J. M. D. Coey, and M. Marezio, *Heat Capacity and Metal-Insulator Transitions in Ti_4O_7 Single Crystals*, Physical Review Letters **32**, 1318 (1974).

[111] S. Lakkis, C. Schlenker, B. K. Chakraverty, R. Buder, and M. Marezio, *Metal-Insulator transitions in Ti_4O_7 single crystals: Crystal characterization, specific heat, and electron paramagnetic resonance*, Physical Review B **14**, 1429 (1976).

[112] L. N. Mulay and W. J. Danley, *Cooperative Magnetic Transitions in the Titanium–Oxygen System: A New Approach*, Journal of Applied Physics **41**, 877 (1970).

[113] C. Acha, M. Monteverde, M. Núñez-Reguiero, A. Kuhn, M. A. Alario Franco, *Electrical resistivity of the Ti_4O_7 Magneli phase under high pressure*, arXiv:condmat/0310048.

[114] M. Marezio and P. D. Dernier, *The Crystal Structure of Ti_4O_7, a Member of the Homologous Series Ti_nO_{2n-1}*, Journal of Solid State Chemistry **3**, 340 (1971).

[115] M. Marezio, D. B. McWhan, P. D. Dernier, and J. P. Remeika, *Structural Aspects of the Metal-Insulator Transition in Ti_4O_7*, Journal of Solid State Chemistry **6**, 213 (1973).

[116] M. Marezio, D. Tranqui, S. Lakkis, and C. Schlenker, *Phase transitions in Ti_5O_9 single crystals: Electrical conductivity, magnetic susceptibility, specific heat, electron paramagnetic resonance, and structural aspects*, Physical Review B **16**, 2811 (1977).

[117] Y. Le Page and P. Strobel, *Structural Chemistry of the Magnéli Phases Ti_nO_{2n-1} ($4 \leq n \leq 9$). – II. Refinements and Structural Discussions*, Journal of Solid State Chemistry **44**, 273 (1982).

[118] Y. Le Page and P. Strobel, *Structural Chemistry of the Magnéli Phases Ti_nO_{2n-1} ($4 \leq n \leq 9$). – III. Valence Ordering of Titanium in Ti_6O_{11} at 130 K*, Journal of Solid State Chemistry **47**, 6 (1983).

[119] C. Schlenker and M. Marezio, *The order-disorder transition of Ti^{3+}-Ti^{3+} pairs in Ti_4O_7 and $(Ti_{1-x}V_x)_4O_7$*, Philosophical Magazine B **42**, 453 (1980).

[120] Y. Le Page and M. Marezio, *Structural Chemistry of the Magnéli Phases Ti_nO_{2n-1} ($4 \leq n \leq 9$). – IV. Superstructure in Ti_4O_7 at 140 K*, Journal of Solid State Chemistry **53**, 13 (1984).

[121] M. Abbate, R. Potze, G. A. Sawatzky, C. Schlenker, H. J. Lin, L. H. Tjeng, C. T. Chen, D. Teehan, and T. S. Turner, *Changes in the electronic structure of Ti_4O_7 across the semiconductor–semiconductor-metal transitions*, Physical Review B **51**, 10150 (1995).

[122] K. Kobayashi, T. Susaki, A. Fujimori, T. Tonogai, and H. Takagi, *Disorder effects in the bipolaron system Ti_4O_7 studied by photoemission spectroscopy*, Europhysics Letters **59**, 868 (2002).

[123] L. L. Van Zandt, J. M. Honig, and J. B. Goodenough, *Resistivity and Magnetic Order in Ti_2O_3*, Journal of Applied Physics **39**, 594 (1968).

[124] L. F. Mattheiss, *Electronic structure of rhombohedral Ti_2O_3*, Journal of Physics: Condensed Matter **8**, 5987 (1996).

[125] A. I. Poteryaev, A. I. Lichtenstein, and G. Kotliar, *Non-local Coulomb interactions and metal-insulator transition in Ti_2O_3: a cluster LDA+DMFT approach*, arXiv: cond-mat/0311319.

[126] C. E. Rice and W. R. Robinson, *High-Temperature Crystal Chemistry of Ti_2O_3: Structural Changes Accompanying the Semiconductor–Metal Transition*, Acta Crystallographica **B33**, 1342 (1977).

[127] U. Schwingenschlögl, V. Eyert, and U. Eckern, *Octahedral tilting in $ACu_3Ru_4O_{12}$ (A=Na,Ca,Sr,La,Nd)*, Chemical Physics Letters **370**, 719 (2003).

[128] P. M. Woodward, *Octahedral Tilting in Perovskites. – I. Geometrical Considerations*, Acta Crystallographica **B53**, 32 (1997).

[129] P. M. Woodward, *Octahedral Tilting in Perovskites. – II. Structure Stabilizing Forces*, Acta Crystallographica **B53**, 44 (1997).

[130] Z. Zeng, M. Greenblatt, M. A. Subramanian, and M. Croft, *Large Low-Field Magnetoresistance in Perovskite-type $CaCu_3Mn_4O_{12}$ without Double Exchange*, Physical Review Letters **82**, 3164 (1999).

[131] M. A. Subramanian, D. Li, N. Duan, B. A. Reisner, and A. W. Sleight, *High Dielectric Constant in $ACu_3Ti_4O_{12}$ and $ACu_3Ti_3FeO_{12}$ Phases*, Journal of Solid State Chemistry **151**, 323 (2000).

[132] C. C. Homes, T. Vogt, S. M. Shapiro, S. Wakimoto, and A. P. Ramirez, *Optical Response of High-Dielectric-Constant Perovskite-Related Oxide*, Science **293**, 673 (2001).

[133] M. A. Subramanian and A. W. Sleight, *$ACu_3Ti_4O_{12}$ and $ACu_3Ru_4O_{12}$ perovskites: high dielectric constants and valence degeneracy*, Solid State Sciences **4**, 347 (2002).

[134] M. Labeau, B. Bochu, J. C. Joubert, and J. Chenavas, *Synthèse et caractérisation cristallographique et physique d'une série de composés $ACu_3Ru_4O_{12}$ de type perovskite*, Journal of Solid State Chemistry **33**, 257 (1980).

[135] S. G. Ebbinghaus, A. Weidenkaff, and R. J. Cava, *Structural Investigations of $ACu_3Ru_4O_{12}$ (A=Na,Ca,Sr,La,Nd) – A Comparison between XRD-Rietveld and EXAFS Results*, Journal of Solid State Chemistry **167**, 126 (2002).

[136] I. D. Brown and D. Altermatt, *Bond-valence parameters obtained from a systematic analysis of the Inorganic Crystal Structure Database*, Acta Crystallographica **B41**, 244 (1985).

[137] V. Eyert, C. Laschinger, T. Kopp, and R. Frésard, *Extended moment formation and magnetic ordering in the trigonal chain compound $Ca_3Co_2O_6$*, Chemical Physics Letters **385**, 249 (2004).

[138] K. E. Stitzer, J. Darriet, and H. C. zur Loye, *Advances in the synthesis and structural description of 2H-hexagonal perovskite-related oxides*, Current Opinion in Solid State and Materials Science **5**, 535 (2001).

[139] S. Niitaka, H. Kageyama, M. Kato, K. Yoshimura, and K. Kosuge, *Synthesis, Crystal Structure, and Magnetic Properties of New One-Dimensional Oxides Ca_3CoRhO_6 and Ca_3FeRhO_6*, Journal of Solid State Chemistry **146**, 137 (1999).

[140] M. J. Davis, M. D. Smith, and H.-C. zur Loye, *Crystal growth, structural characterization and magnetic properties of Ca_3CuRhO_6, $Ca_3Co_{1.34}Rh_{0.66}O_6$, and Ca_3FeRhO_6*, Journal of Solid State Chemistry **173**, 122 (2003).

[141] S. Niitaka, K. Yoshimura, K. Kosuge, M. Nishi, and K. Kakurai, *Partially Disordered Antiferromagnetic Phase in Ca_3CoRhO_6*, Physical Review Letters **87**, 177202 (2001).

[142] A. Maignan, C. Michel, A. C. Masset, C. Martin, and B. Raveau, *Single crystal study of the one dimensional $Ca_3Co_2O_6$ compound: five stable configurations for the Ising triangular lattice*, European Physical Journal B **15**, 657 (2000).

[143] S. Niitaka, K. Yoshimura, K. Kosuge, A. Mitsuda, H. Mitamura, and T. Goto, *Pulsed high field study of new partially disordered antiferromagnetic phase in* Ca_3CoRhO_6, Journal of Physics and Chemistry of Solids **63**, 999 (2002).

[144] S. Niitaka, H. Kageyama, K. Yoshimura, K. Kosuge, S. Kawano, N. Aso, A. Mitsuda, H. Mitamura, and T. Goto, *High-Field Magnetization and Neutron Diffraction Studies of One-Dimensional Compound* Ca_3CoRhO_6, Journal of the Physical Society of Japan **70**, 1222 (2001).

[145] B. Raquet, M. N. Baibich, J. M. Broto, H. Rakoto, S. Lambert, and A. Maignan, *Hopping conductivity in one-dimensional* $Ca_3Co_2O_6$ *single crystals*, Physical Review B **65**, 104442 (2002).

[146] E. V. Sampathkumaran and A. Niazi, *Superparamagnetic-like ac susceptibility behavior in the partially disordered antiferromagnetic compound* Ca_3CoRhO_6, Physical Review B **65**, 180401 (2002).

[147] M. Loewenhaupt, W. Schäfer, A. Niazi, and E. V. Sampathkumaran, *Evidence for the coexistence of low-dimensional magnetism and long-range order in* Ca_3CoRhO_6, Europhysics Letters **63**, 374 (2003).

[148] C. Laschinger, T. Kopp, V. Eyert, and R. Frésard, *Microscopic Deviation of Magnetic Coupling in* $Ca_3Co_2O_6$, Journal of Magnetism and Magnetic Materials, in press, arXiv:cond-mat/0310500.

[149] M. H. Whangbo, D. Dai, H. J. Koo, and S. Jobic, *Investigations of the oxidation states and spin distributions in* $Ca_3Co_2O_6$ *and* Ca_3CoRhO_6 *by spin-polarized electronic band structure calculations*, Solid State Communications **125**, 413 (2003).

[150] R. Frésard, C. Laschinger, T. Kopp, and V. Eyert, *The Origin of Magnetic Interactions in* $Ca_3Co_2O_6$, Physical Review B, in press, arXiv:cond-mat/0309031.

[151] V. Eyert, K.-H. Höck, and P. S. Riseborough, *The Electronic Structure of* La_2BaCuO_5*: A Magnetic Insulator*, Europhysics Letters **31**, 385 (1995).

[152] R. Weht and W. E. Pickett, *Extended Moment Formation and Second Neighbor Coupling in* Li_2CuO_2, Physical Review Letters **81**, 2502 (1998).

[153] M. Methfessel and A. T. Paxton, *High-precision sampling for Brillouin-zone integration in metals*, Physical Review B **40**, 3616 (1989).

[154] S. Niitaka, K. Yoshimura, K. Kosuge, K. Mibu, H. Mitamura, and T. Goto, *Magnetic and* ^{57}Fe *Mössbauer studies of* Ca_3FeRhO_6, Journal of Magnetism and Magnetic Materials **260**, 48 (2003).

Figures and Tables

Index

Acknowledgements

I would like to express my gratitude to everyone who contributed to the successful completion of this thesis. In particular, I am much obliged to

- Priv.-Doz. Dr. V. Eyert for introducing me to a most interesting field of research as well as for innumerable fruitful discussions.
- Prof. Dr. U. Eckern for supporting my work over the last three years.
- Prof. Dr. T. Kopp for his cooperation and for providing the second report.
- Prof. Dr. K. Schwarz for providing the third report.
- Priv.-Doz. Dr. P. Schwab for his continuous interest and many helpful discussions.
- Dr. M. Gruber, C. Laschinger, Dr. T. Lück, and Dr. C. Schuster for a lot of interesting suggestions.
- C. Wunsch for proofreading the English of this thesis.
- All colleagues of the physics institute for a most pleasant atmosphere.

Curriculum Vitae

9. Februar 1977	Geboren in Bobingen
1983 bis 1987	Besuch der Grundschule Augsburg Vor dem Roten Tor
1987 bis 1996	Besuch des Rudolf-Diesel-Gymnasiums Augsburg
28. Juni 1996	Abitur
1996 bis 2001	Studium der Physik an der Universität Augsburg Diplomarbeit bei Prof. Dr. R. Claessen mit dem Thema: *Photoemissionsspektroskopie an quasi-eindimensionalen organischen Metallen*
28. Juni 2001	Diplom in Physik
Seit Juli 2001	Wissenschaftlicher Angestellter, Universität Augsburg, Lehrstuhl für Theoretische Physik II (Prof. Dr. U. Eckern)